BIOBASED LUBRICANTS AND GREASES

Tribology Series

Martin & Ohmae	Nanolubricants	April 2008
Khonsari & Booser	Applied Tribology – Bearing Design and Lubrication, 2nd Edition	April 2008
Stachowiak (ed)	Wear – Materials, Mechanisms and Practice	November 2005
Lansdown	Lubrication and Lubricant Selection, 3rd Edition	November 2003
Neale, Polak & Priest (eds)	Handbook of Surface Treatment and Coatings	May 2003
Sherrington, Rowe & Wood (eds)	Total Tribology – towards an integrated approach	December 2002
Kragelsky, Alisin, Myshkin & Petrokovets (eds)	Tribology – Lubrication Friction and Wear	April 2001
Stolarski & Tobe	Rolling Contacts	December 2000
Neale and Gee	Guide to Wear Problems and Testing for Industry	October 2000

BIOBASED LUBRICANTS AND GREASES

TECHNOLOGY AND PRODUCTS

Lou A.T. Honary
Professor and Founding Director
National Ag-Based Lubricants Center
University of Northern Iowa, USA

Erwin Richter
National Ag-Based Lubricants Center
University of Northern Iowa, USA

A John Wiley and Sons, Ltd, Publication

This edition first published 2011

Registered office
John Wiley & Sons Ltd, The Atrium, Southern Gate, Chichester, West Sussex, PO19 8SQ, United Kingdom

For details of our global editorial offices, for customer services and for information about how to apply for permission to reuse the copyright material in this book please see our website at www.wiley.com.

Library of Congress Cataloguing-in-Publication Data

Honary, Lou A. T.
Biobased lubricants and greases : chemistry, technology, and products / Lou A.T. Honary, Erwin Richter.
p. cm.
Includes index.
ISBN 978-0-470-74158-0 (cloth)
1. Vegetable oils as fuel. 2. Lubricating oils–Biocompatibility. 3. Plant biotechnology. 4. Biomass energy. I. Richter, Erwin William, 1934- II. Title.
TP359.V44H66 2010
665.5'385–dc22 2010025738

A catalogue record for this book is available from the British Library.

Print ISBN: 9780470741580
ePDF ISBN: 9780470971963
oBook ISBN: 9780470971956

Set in 10/12 Times by Thomson Digital, Noida, India
Printed and bound in Singapore by Markono Print Media Pte Ltd

Contents

About the Authors

Lou A.T. Honary

Lou A.T. Honary is a professor and founding director of the University of Northern Iowa's National Ag-Based Lubricants Center. He joined the UNI faculty in 1982 and in 1991, he initiated the research and development of soybean oil based lubricants and greases, leading to the creation of the UNI-NABL Center which is a premier applied research center specifically focused on biobased lubricants and greases.

As an applied researcher, Honary's work has resulted in eight patents or co-patents, two more patents pending and numerous publications and presentations at national and international conferences. With the University of Northern Iowa's Research Foundation, in 2000, Honary formed a commercial lubricants and grease manufacturing company that has brought to the market various biobased products and is recognized as a leader in biobased grease manufacturing. With over 40 commercial products currently on the market having their origins to his research, Honary is considered perhaps the most knowledgeable expert in the area of biobased lubricants and greases in the United States.

Professor Honary has served in leadership capacities in many organizations including memberships in the American Oil Chemist Society (AOCS), Society of Tribologists and Lubrication Engineers (STLE), Society of Automotive Engineers (SAE), National Lubricating Grease Lubricating Institute (NLGI), European Lubricating Grease Institute (ELGI), American Society for Testing and Materials (ASTM), and National Fluid Power Association (NFPA). He has served as president of the Fluid Power Society (FPS), in Iowa, Member At Large on the Board of Directors of the International Fluids Power Association (IFPS), an Officer on the ASTM D02 Committee, and a member of the Board of Directors of NLGI, and chairman of a working group on the performance of biobased greases for ELGI.

Dr. Honary is an entrepreneurial professor recognized for his visionary approach to research. He has served on an Iowa Governor's committees, on a congressionally mandated Biomass Research and Development Advisory Committee under the United States Department of Energy (DOE) and Department of Agriculture (USDA), and has served as a consultant to government and industry. Honary's consulting work has included the preparation of a series of protocols for the creation of specifications for eleven biobased hydraulic oils for the US Department of the Navy to be used as a substitute for conventional hydraulic fluids.

As a passionate promoter of biobased products, Honary is known for many firsts, including patenting the first soybean oil based tractor hydraulic fluid, the first soybean oil based transformer oil, the first soybean oil based wood preservative as a creosote substitute, the first biobased solid stick lubricant for railroads, the first soybean oil based rail curve grease among many other products. In 2010 he and his team introduced a revolutionary efficient and safe heating process using microwave energy for the manufacturing of biobased greases.

Honary is a sought after speaker at various technical conferences owing to his ability to present complex concepts associated with biobased lubricants in a practical and easy to understand approach. This book provides an example of his passion for teaching by presenting an engaging and easy to follow approach, making this book both enjoyable to read and a resource to keep.

Erwin W. Richter

Erwin Richter was born in 1934. He taught in the public schools of Michigan before receiving his PhD in biochemistry from the University of Iowa in 1970. He was a member of the faculty at the University of Northern Iowa from 1963 to 1996. In 2001 he began working at The University of Northern Iowa's National Ag-Based Lubricants Center as a consultant and continues there today. His interests in chemistry led him to develop his knowledge in the area of biolubricant development and testing. He is the author of several books and laboratory manuals dealing with chemistry education.

Preface

Vegetable oils present properties that are suitable for industrial and automotive lubricants and grease applications. They also present potential for usability as an alternative to petroleum when the demand for this finite resource is ever increasing.

The goals of writing this book include educating the next generation of students and professionals in this promising field to create and use biobased lubricants and greases. As the world petroleum resources continue to deplete, resource-poor and developing countries will have to struggle to compete to acquire high priced petroleum and petroleum products. Creating lubricant (and fuel) products from renewable sources can offer self sufficiency and potentially economical alternatives to the countries most desperate for these advantages.

After nearly two decades of research and development of biobased lubricants and greases, we hope that sharing our knowledge and expertise will help to create a long standing resource for the future. Both authors are seasoned professors and researchers, and the book is written in a way that it teaches the concepts for general audience comprehension. For more advanced concepts in biobased lubricants and greases, there are other sources that delve into the engineering and agronomical aspects of our work in greater detail.

With the hope of a better, greener future, we offer this book to our future generations.

Lou Honary
Erwin Richter

Preface

Series Preface

There are increasing concerns and growing regulations over contamination and environmental pollution. One of the major concerns is pollution caused by mineral oils. As the world oil reserves are dwindling the pressures for finding alternative replacements are increasing. Vegetable oils are a biodegradable and renewable source of lubricants. Thus, they seem to be attractive candidates for the replacement of mineral oils. There are not many books on the market comprehensibly describing plant-based lubricants. Therefore, this book is a much welcomed edition to the Tribology Series. The major industrial crop sources of vegetable oils, together with their chemical and physical characteristics, are discussed in this book. The strength of this book is that it not only provides the comprehensive overview of oil producing plants but it also provides the details of oil testing and characterization methods. Potential applications of the vegetable oils are also shown. In this book, apart from vegetable oils, bio-based greases, together with their specifications, are also described. Numerous detailed illustrations are very helpful in conveying the technical information to the reader. This book is richly illustrated with the colour pictures of oil producing plants, their seeds and equipment used in the oil testing and evaluation. Information is provided in an easily accessible form.

This book is a useful source of practical information on vegetable-based lubricants, their characteristics and testing methods. It is recommended for the readers working on plant-based lubricants and tribology. This book would be a useful reference to engineers, technicians and both under and postgraduate students with interest in plant-based lubricants.

Gwidon Stachowiak
University of Western Australia
Perth, Australia

Acknowledgements

I would like to acknowledge the resources and support provided by the University of Northern Iowa's National Ag-Based Lubricants Center. Without access to a wealth of information on the development of biobased lubricants and products it would have been difficult to compile such an array of experiential information for this book.

Also, the staff at UNI-NABL Center were patient during the long hours of writing and were supportive of the efforts in any way possible. Special thanks are due to Meghan K. Reynolds who took the task of proofing some of the technical chapters for consistency and accuracy.

Additionally, special thanks are due to Mike Jensen and Alan Burgess of Environmental Lubricants Manufacturing, Inc. who selflessly shared hands-on and experiential information on their years of experience in commercially marketing biobased lubricants and greases. Their real life experiences in dealing with the post production and performance issues of biobased lubricants and greases have been helpful in pointing out the possible pitfalls of these new products.

My daughter, Shereena R. Honary deserves special thanks for proofreading the entire document for consistency and content. She was able to 'remove Lou Honary's accent' from most of the chapters. Lastly, but certainly not the least importantly, my wife Carol deserves heartfelt thanks for her patience while I spend hours researching and writing the materials for this book.

I am also honored that my colleague, Dr. Erwin Richter was able to contribute a chapter to this book and adding his expertise in chemistry and the benefit of his years of teaching the subject. With his contribution, this book has become much more valuable.

Lou A.T. Honary
Professor and Founding Director
National Ag-Based Lubricants Center
University of Northern Iowa

Summary

This book includes the historical perspective of vegetable oils, their chemical overview, references and data on vegetable oil characteristics in comparison to petroleum oils, associated breeding and genetic modification technologies, and the successful products currently on the market. The book is written for the generalist and provides sufficient information in each area but is not too detailed as desired by the specialist. Illustrations and pictorial descriptions are utilized to help with conveyance of technical details. The book is unique in covering a combination of biobased lubricants and greases. This book should be a useful resource for a wide array of readers interested in understanding the potential of biobased lubricants.

Introduction

In the United States, national policy has created a new impetus for promoting and using biobased products. Moreover, worldwide demand for petroleum combined with geopolitical issues related to petroleum producing regions has created a more accepting market for alternative lubricants and greases.

The United States is the largest worldwide supplier of agricultural products and oilseeds like soybeans. Advanced mechanized farming, combined with superior seed and related technologies, has made the country's farmers highly efficient producers of commodity crops. Similarly, Canadian and European countries have invested significant amounts of resources in advancing production technologies for canola and rapeseed respectively. Other countries like Brazil have also increased production of oilseed crops like soybeans to the point of being competitive in the world market.

Since the authors' research work has been mostly on the US-based crop oils, primarily soybean oil, the majority of the examples used in this book deal with soy oil. In industrial countries like the United States, an advanced and capable infrastructure for producing and shipping agricultural products has for many years resulted in surpluses far beyond the needs of their citizens. As a result, for years when petroleum prices were cheap, farmers in the United States, Canada, and Europe had to perpetually seek out new uses and new markets for their surplus agricultural products.

The authors performed their initial research at the University of Northern Iowa's National Ag-Based Lubricants (NABL) Center funded by an association of the state of Iowa soybean growers. The goal of this research was to find new uses for the perennial surplus soybeans and soybean oil in non-food areas. As the worldwide demand for petroleum continued to increase, and petroleum prices skyrocketed, the cost parity for biofuels and bioproducts became a reality.

Materials presented in this book are derived from actual evaluation of properties of various vegetable oils and their performance as lubricants. In so far as is possible their performance properties are compared to their petroleum-based counterparts. The information presented includes data on the performance of vegetable oils including canola (and rapeseed) oil, sunflower oil, palm oil and some mixtures of these oils.

1

Historical Development of Vegetable Oil-based Lubricants

1.1 Introduction

Lubrication has probably been known to humans since the invention of the wheel. Recorded pictorial documents point to the use of water or edible oils as a lubricity liquid used by Egyptian pyramid builders when rolling large pieces of rocks on wooden rollers. Figure 1.1 shows a painting from an inner wall of the Tehuti-Hetep tomb. In this painting one worker is depicted as pouring a liquid in front of the rollers while others are pulling on the load [1].

The use of animal fat or vegetable oils for lubricating the wheels of horse-drawn carts and carriages has been well documented. It is not hard to visualize that the use of wooden axles and wheels, or even a combination of metallic wheels and axles would create friction and wear. Lubrication then becomes instinctive for human-made mechanical machinery, always seeking the most stable and efficient lubricant. Examples of these mechanical machines include Persian carriages, wind turbines running in wooden shafts, and waterwheels, all requiring some form of lubrication (Figure 1.2).

When we concern ourselves with actions or processes that help reduce friction we are engaged in tribology (tri-ball-ogy). The word tribology is used esoterically in lubrication and engineering sciences but is somewhat unknown outside these fields. Tribology is the science related to friction and wear or of reducing friction and wear. *Tribo*, is a Greek root meaning "rub" or friction combined with the word *logos* meaning "related" or, "the logic of." The Society of Tribologists and Lubrication Engineers (STLE) has a large membership in countries around the world. Many other technical and engineering societies devote a section to tribology-related subjects.

The industrial revolution saw drastically higher demands for lubricants to both lubricate moving parts and for energy transfer, like in hydraulic fluids. The hydraulic fluid acts as an energy transfer medium in hydraulic piping and at the same time helps to lubricate, seal, and remove heat and contaminants from components.

Biobased Lubricants and Greases: Technology and Products, First Edition. Lou A.T. Honary and Erwin Richter

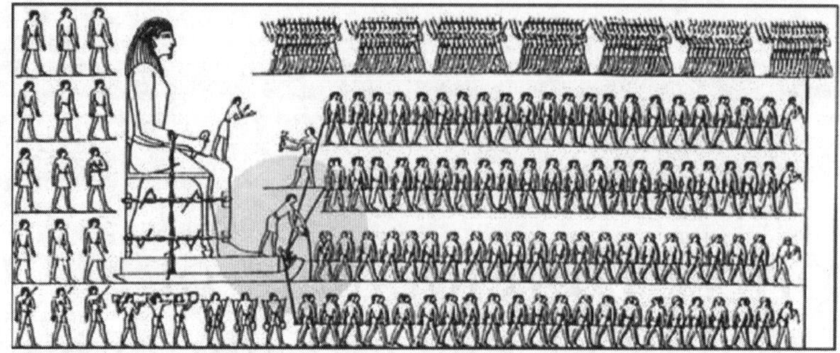

Figure 1.1 Egyptian pyramid builder applying oil on rollers to reduce friction [1]. Reproduced with permission from Dowson, D., *History of Tribology*, 1998, John Wiley & Sons, Ltd, Chichester

Joseph Bramah (1749–1814) is known for a number of inventions including the hydraulic press. In 1795, Bramah invented the hydraulic press capable of exerting huge forces for forming metals. Bramah applied the known hydraulic principle of Pascal's Law to a practical application of gaining mechanical advantage by the use of cylinders. His press offered a real and practical application of Pascal's Law to manufacturers that had to otherwise rely on often less efficient bulky mechanical systems. Applications of Bramah's invention include the hydraulic car lift, presses for forming metals, and the hydraulic brakes. Bramah's press is considered to have been a major contributor toward the advances made during the Industrial Revolution (Figure 1.3).

The need for more stable oils concurrent with the need for better sealing materials for high pressure hydraulic cylinders in presses led to the development of oils and the use of chemical additives. Initial seals used in hydraulic cylinders were made of leather, which would absorb the lubricant, causing it to swell and seal but also needing more frequent replacement than the current elastomeric seals. Water was one of the initial fluids used for hydraulic cylinders.

The automobile's brake system is a good example of how hydraulic oil transfers the force of the driver's foot to the brake pads to stop the vehicle. In this case, the main purpose of the brake fluid is energy transfer, although it also lubricates and seals the pistons in the master and wheel cylinders (Figure 1.4).

Figure 1.2 Persian war chariots and windmills used animal fat or vegetable oils for lubrication. Courtesy of Grenada Studios, St Petersburg, Russia

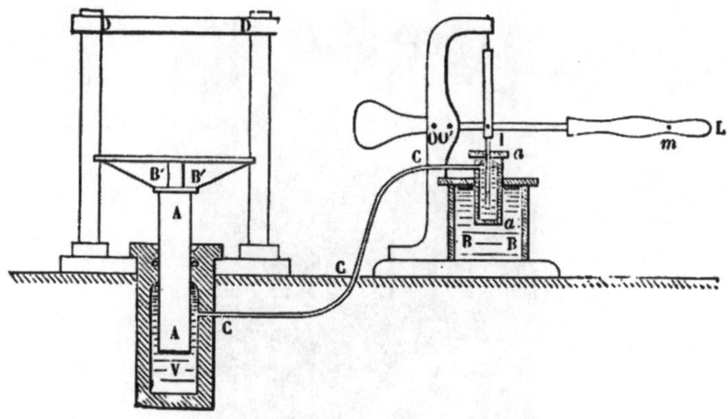

Figure 1.3 Schematic of Bramah's Press

1.2 Pioneering Industrial Uses of Vegetable Oils

The idea of using vegetable-based products as lubricant or fuel is not new. For example, Rudolf Diesel used peanut oil to power one of his diesel engines in 1900 during a power show in Paris. Technology in the use of biolubricants and fuels was put on hold however, due to the abundance and low cost of petroleum.

When investigating the industrial uses of vegetable oils in the United States, two names stand out. The industrialist Henry Ford I and Dr George Washington Carver (Figure 1.5), who was a pioneering agricultural researcher! Henry Ford I had a vision of using crop-based materials in making cars and tractors and creating a closed circle of cradle to grave renewable products.

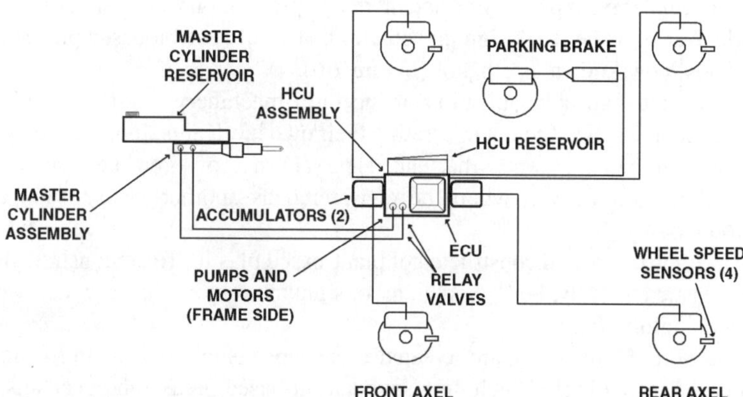

Figure 1.4 A schematic of an automobile brake system with hydraulic lines using oil as the energy transfer medium

Figure 1.5 Henry Ford (right) and George Washington Carver. Courtesy of The Henry Ford

George Washington Carver's research resulted in over 500 products from peanuts, sweet potatoes, and pecans. He graduated in 1894 from Iowa State College of Agriculture and Mechanic Arts (now Iowa State University, in Ames, Iowa). Carver joined the faculty of the college and continued his studies in bacteriological laboratory work in systematic botany. Later he joined the faculty and assumed the directorship of the Department of Agricultural Research at Tuskegee Normal (now Tuskegee University) in 1896. Henry Ford has been credited for leading the development of biobased products in the early twentieth century. He also worked closely with George Washington Carver, recognizing their mutual vision. They shared a vision of a future in which agricultural products would be put to new uses to create products and industries.

Henry Ford believed that agriculture could supply the industry with renewable raw materials. His invention of the Ford Model T was designed as an economical vehicle for the masses. The concept of mass production and increased production were also promoted by Ford. During World War II, he began designing vehicles that would use biobased plastic bodies and corn-based fuels, now known as ethanol (Figure 1.6).

Paying attention to the mechanization of the farm machinery, Ford was convinced that farmers could become self sufficient in creating their own lubricants, food and fuels from their farm renewable products. Towards that end, Henry Ford sponsored the research activities of George Washington Carver, whom he considered as another visionary in the use of biobased products.

The car was lightweight and constructed of plant-based plastic. To demonstrate the stability of a flexible plastic car body, Ford used an ax to simulate minor crashes which would result in the ax bouncing off (Figure 1.7).

In 2004 the Ford Motor Company created a concept vehicle that included many of the existing biobased technologies, including soybean oil-based grease and gear oils. This was appropriately called Model U as a follow up to the popular model T that was created by the founder of the company decades earlier (Figure 1.8). The authors provided biobased grease for use in this concept vehicle.

Figure 1.6 A "soybean car" containing biobased materials. Courtesy of The Henry Ford

1.3 Petroleum

Petroleums from different regions of the world have different properties but in general, petroleum is decayed plant and animal remains with the main constituents being the same regardless of the source. In the early twentieth century United States, the discovery of petroleum in Spindletop, Texas (Figure 1.9) created a rush for drilling, which resulted in the availability of cheap crude oil. The Gladys City Oil, Gas, and Manufacturing Company, was the first oil company to drill in Spindletop. In its first year, Spindletop production reached 3.59 million barrels of oil. By the second year, production had reached 17.4 million barrels.

Figure 1.7 Henry Ford demonstrating the strength of the bioplastic car body. Courtesy of The Henry Ford

Figure 1.8 Ford Model U utilizing biobased materials where possible. Photo courtesy of the Ford Motor Company

Eventually production reached nearly 100,000 barrels per day leading to a major economic boom for the area [2].

With the high-energy intensity of petroleum as a motor vehicle fuel, no other automotive power plant (like steam engines and electrically driven motors for vehicles), nor any other base oils (like vegetable oils) had a chance to compete. The cheap petroleum oil was a blessing for the growth of the fledgling automotive industry and for the use of gasoline byproducts, which are heavier hydrocarbons, for lubricants and greases. At the same time, however, it was

Figure 1.9 Discovery of oil: Spindletop, Texas led to popularity of gasoline as fuel and mineral oils for lubrication

a curse for creating an ever-increasing demand for petroleum at the expense of all other alternatives. This included electric and steam power plants for vehicles, and animal fat or vegetable oils for lubricants. Only during the two World Wars did petroleum supply interruptions result in a shift in attention towards the use of vegetable oils for fuels as biodiesel and lubricants for machinery or hydraulic energy transfer.

The advances made in extraction and refining of petroleum resulted in the creation of fuels, oils, and chemicals to complement each other. The single viscosity engine oils, for example, required the use of thinner oil in the winter season to be replaced by thicker or higher viscosity oils to use in the summer. Later, by addition of polymeric additives, multiviscosity grade oils were created that would work in any temperature.

The rapidly increasing usage of automobiles in the United States along with the invention of mass production techniques by Eli Olds (and utilized by Henry Ford), created a perfect storm for the superiority of petroleum-based lubricants.

A fire in the Eli Olds production facility destroyed his inventory of parts and vehicles, leaving him with only one unit of his then famous "curved dash automobile" (Figure 1.10) that had been stored in his garage. The necessity of having to produce over two-dozen vehicles on order with no production facility led him to contract with various shops in Detroit. These shops in turn built components of his cars to the exact dimension of the one surviving curved dash vehicle. This, in effect, created the concept of mass production, whereby accurately produced components of the car manufactured by different companies would be assembled in one place to complete vehicles. Detroit thus became Motor Town, or Motown, with almost every shop making parts for some car company.

Figure 1.10 Old's curved dash vehicle started the mass assembly trend. Picture of Eli Olds Car advertisement. See Plate 1 for the color figure

Forces that began to turn the tide toward the use of vegetable oils as a substitute for petroleum products included increased concern for the environment, increased demand for petroleum, and geopolitical pestering. Many factors, such as oil spills like the infamous *Exxon Valdez* spill in Alaska's Prince William Sound of March 1989, the BP oil well explosion in the Gulf Mexico in 2010, global warming, and the demand for oil by developing countries, have created this turning point. It is estimated that in the brief period of 100 years since the discovery of oil in Spindletop, we have exhausted half of the world's petroleum.

During the 1980s, European researchers encouraged by the agriculture community in Europe began to explore the use of vegetable oils as hydraulic fluids and other industrial lubricants. The concern for the environment and lobbying of farmers' organizations led to mandates for the use of biodegradable products in certain parts of Europe. For example, in the 1980s, the German government required the use of biodegradable hydraulic oils in Black Forest regions. During this period, the European community created environmental seals and emblems to identify the "Environmentally Aware" lubricants.

By the 1990s, many North American companies began to follow the Europeans' leads on creating biodegradable products. In the 1990s' conferences held by the American Society for Testing and Materials, over 40 North American companies had representatives in attendance to discuss their efforts in creating biodegradable products and help in establishing standards.

In the United States, for example, The Lubrizol Corporation invested significant amounts of resources in creating additive packages for vegetable oil (specifically sunflower oil) based lubricants. The list of additive packages and products from Lubrizol was comprehensive and included food grade products, two-stroke-cycle engine oils, and universal tractor hydraulic fluids.

Eventually, the relatively low price of petroleum and a lack of mandates for using biodegradable or renewable products diminished the investments in research and development (R&D) for these products. By the late 1990s, the only groups continuing to fund R&D of industrial products and lubricants made from renewable materials were farmers groups like the United Soybean Board, representing US soybean farmers, or the US Department of Agriculture.

In 2000, the United States' Farm Bill, which was a 5-year plan for the advancement of agriculture, included provisions for the promotion and use of biodegradable and renewable products. The US government selected "the leadership by example" approach in an attempt to avoid mandates and allow the free market enterprise to bring about the success of renewable products. This approach required federal agencies to purchase and use biobased products so as to prove viability of performance and eventually lead to commercial success in competitive private sector markets. Biobased and renewable based lubricants now have a presence in the world market and are anticipated to grow in technology and use.

In order to better follow the future of biobased lubricants *vis-à-vis* petroleum lubricants it is important to understand the different plant oils and petroleum oils. Within this context, the basic concepts relating to these two oils will be covered.

References

1. Madius, C. and Smet, W. (2009) *Grease Fundamentals: Covering the Basic of Lubricating Grease*, Axel Christiansson, Heijningen, The Netherlands, www.AXELCH.com.
2. Materials on Spindletop Texas retrieved April 18, 2010: http://www.eoearth.org/article/Spindletop,_Texas.

2

Chemistry of Lubricants

2.1 The Nature of the Carbon Atom

The basic building blocks of all elements in the universe are protons, neutrons, and electrons. Protons have a positive electrical charge, electrons have a negative electrical charge, and neutrons are neutral in electrical charge. The magnitude of the charge on electrons and protons is exactly equal but opposite so that together they neutralize each other. These subatomic particles all have mass. The mass of the proton is roughly 2000 times the mass of the electron. Neutrons have roughly the same mass as that of the proton. These particles are assigned relative mass numbers. Protons are assigned 1 atomic mass unit (amu), neutrons are assigned 1 amu and electrons, since they are so much less massive than the other two subatomic particles, are assigned 0 amu. All atoms are electrically neutral, which means that the number of protons and electrons they contain are equal. The combined mass of the protons and neutrons in an atom make up the atomic mass of an atom. Atomic masses of all the elements can be found on the periodic chart of the elements.

The simplest and most abundant element in the universe is hydrogen. Hydrogen has an electrically positive nucleus consisting of a single proton. An electron orbits around the nucleus creating a neutral hydrogen atom. The property that changes as atoms become more complex is the number of protons and neutrons in the nucleus. In contrast, the most complex naturally occurring atom is uranium with 92 protons and 146 neutrons in its nucleus.

2.2 Carbon and Hydrocarbons

Carbon is a relatively simple atom consisting of six extranuclear electrons, six protons and six, seven, or eight neutrons in its nucleus. Keeping in mind the mass numbers assigned to the particles, carbon atoms can have a mass of 12, 13, or 14 amu. These are referred to as carbon isotopes and are designated C^{12}, C^{13}, and C^{14}. Carbon-14 (C^{14}) is radioactive and is used in radiocarbon dating. Carbon-14 is more concentrated in oilseeds than it is in petroleum. This fact can be useful in determining whether a given fuel or lubricant is biobased. Determination of biobased content for the USDA labeling program is based on the amount of C-14 (carbon

Biobased Lubricants and Greases: Technology and Products, First Edition. Lou A.T. Honary and Erwin Richter

from renewable sources) in a formulated lubricant as opposed to C-12 (carbon from fossilized sources). The nucleus of an atom has very little to do with its chemistry, therefore all the isotopes of carbon behave chemically similar to each other.

To understand the chemistry of an atom we must look at its electrons. Electrons are distributed around the nucleus of an atom in various energy levels called shells. The shell furthest from the nucleus is called the **valence** shell. It is the number of electrons in the valence shell that determines how a given element will react chemically with another element. The number of electrons in the valence shell of atoms varies from one to eight. Carbon has four electrons in its valence shell and therefore has a **valence number** of 4. Elements with eight electrons in the valence shell do not react chemically with any other elements and are therefore referred to as "inert." It is the valence number of an atom that determines how many chemical bonds it can form with other atoms. Atoms form bonds with other atoms by sharing electrons or transferring electrons with each other until each has an outer shell of electrons that resembles the structure of an inert substance. Carbon therefore, can form four chemical bonds with other atoms.

Carbon is somewhat unique among the elements in that it *can form chemical bonds with itself*. This property of carbon allows it to form carbon atom chains of various lengths. This property of carbon occurs in petroleum where carbon is chemically bound to hydrogen and to itself to form hydrocarbon chains. A complex system has evolved to name the myriad chemical compounds that exist due to carbon's ability to bond with itself. It is not the purpose of this book to go into this complex system, however, a basic understanding of the vocabulary and shorthand used to describe certain carbon compounds is necessary in order for reader comprehension of carbon and its compounds.

Petroleum consists of a mixture of hydrocarbon molecules of varying molecular mass. The simplest hydrocarbon molecule is one consisting of a single carbon atom bound to four hydrogen atoms. The compound is referred to as methane and is the major component of natural gas. It has the *empirical* chemical formula CH_4. More information is gained, however, by looking at its *structural* formula (Figure 2.1).

One can easily see from this formula that carbon is bound to four hydrogen atoms. The lines that connect hydrogen to carbon are referred to as single chemical bonds and represent a single ***pair*** of electrons.

Ethane is a hydrocarbon where carbon bonds to itself to create a two-carbon atom chain. It has the empirical formula C_2H_6. Its structural formula is shown in Figure 2.2.

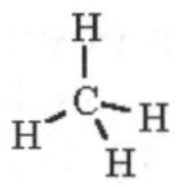

H
C
H H H

Figure 2.1 Structural formula for methane

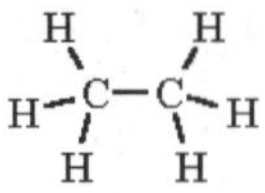

H H
H C C H
H H

Figure 2.2 Structural formula for ethane

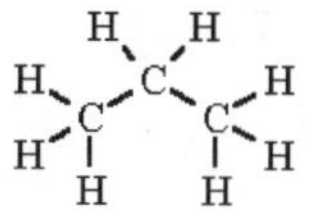

Figure 2.3 Structural formula for propane

A hydrocarbon consisting of a chain of three carbon atoms is referred to as propane (C_3H_8) and has the structural formula shown in Figure 2.3.

The above compounds are all compounds of carbon and hydrogen. They differ in complexity and are named differently. In order to become familiar with carbon chemistry it is important that the reader become familiar with the systematic naming system used to describe carbon compounds. The International Union of Pure and Applied Chemistry (**IUPAC) has adopted a naming system based on the number of carbon atoms in a given molecule as well the atoms to which carbon is chemically bonded**. Common elements other than hydrogen being bonded to carbon include oxygen, nitrogen, and sulfur.

The first ten hydrocarbons in petroleum are named as follows:

CH_4	Methane
C_2H_6	Ethane
C_3H_8	Propane
C_4H_{10}	Butane
C_5H_{12}	Pentane
C_6H_{14}	Hexane
C_7H_{16}	Heptane
C_8H_{18}	Octane
C_9H_{20}	Nonane
$C_{10}H_{22}$	Decane

Notice that the name of all of these compounds ends with the suffix **"ane." The suffix of these names deal with a property known as "saturation**." The reason for this nomenclature is that carbon has the ability to share more than one valence electron with another carbon atom. When this happens, the compound is said to be "unstaturated." **An unsaturated carbon compound will end with either "ene" or "yne" as the suffix**. This will be dealt with later in this chapter.

The following examples will illustrate how some of the IUPAC nomenclature is used.

C_3H_8 is the empirical formula for propane. The "pro" part of the name propane is a carbon atom count and indicates three carbon atoms. The "ane" part of the name tells us that the compound is saturated, which is to say that there are no carbon atoms present in the molecule that share more than one electron with another carbon atom. The structural formula for propane is shown in Figure 2.4.

C_3H_6 is the empirical formula for propene. **The "pro" part of the name again tells us that there are three carbon atoms in the molecule**. The **suffix "ene" tells us that there is a double bond(two shared pairs of electrons) between two of the carbon atoms**. The presence of a double bond makes the compound unsaturated. The structural formula for propene is shown in Figure 2.5.

```
     H    H    H
     |    |    |
H -- C -- C -- C -- H
     |    |    |
     H    H    H
```

Figure 2.4 Propane

```
H        H    H
 \       |    |
  C  ==  C -- C -- H
 /            |
H             H
```

Figure 2.5 Propene

C_3H_4 is the empirical formula for propyne. Again, "pro" indicates three carbon atoms. The "yne" suffix indicates that two carbon atoms share three pairs of electrons between them to form a triple bond. The structural formula for propyne is shown in Figure 2.6.

At this point, the reader might try his or her skill at naming a hydrocarbon. The empirical formula for this particular hydrocarbon is C_5H_{10}. See if you can name the compound in whose structural formula is shown in Figure 2.7.

Did you name it correctly? "**Pent**" must be part of the name since there are **five carbon atoms** involved. "**Ene**" must also be part of the name since **there is a double bond** between two of the carbon atoms. Thus the name "pentene" would describe the molecule pictured.

As the length of the carbon chain of a molecule increases, so do its chemical and physical properties. **In petroleum, the hydrocarbons methane, ethane, propane and butane are gases at room temperature**. Propane and butane are easily compressed to liquids. Liquid propane (LP) is used for home heating and grain drying. Butane is compressed to a liquid to make small hand-held lighters. Pentane and hexane are low boiling point liquids and make excellent solvents and degreasers. Heptane, octane, and nonane, when mixed together with other substances, make gasoline. Longer chain hydrocarbons make up diesel fuel, home

```
 H
  \
H--C-C≡C-H
  /
 H
```

Figure 2.6 Propyne

```
H       H    H    H    H
 \      |    |    |    |
  C  == C -- C -- C -- C -- H
 /           |    |    |
H            H    H    H
```

Figure 2.7 Exercise formula

C_xH_y	+	O_2	→	CO_2	+	H_2O	+	Heat
Hydrocarbon		Oxygen		Carbon dioxide		Water		

Figure 2.8 General chemical reaction to generate heat

heating oil, lightweight engine oils, greases, and other such substances. The longest carbon chain molecules in petroleum become solids when placed at room temperature. These solids would be substances such as waxes or tar. Petroleum is a mixture of all of these hydrocarbons. It must be separated into its components in order for it to be a valuable commodity. Society has progressed to the point it is today largely due to the chemical energy contained in petroleum.

Most petroleum is used today in combustion reactions to produce heat. In a combustion reaction, atmospheric oxygen is combined with the hydrogen and carbon in hydrocarbons to produce carbon dioxide and water. The general reaction for the generation of heat is shown in Figure 2.8.

The heat produced in this reaction drives electrical power plants and machines like automobiles, airplanes, lawn mowers, farm tractors, and the like. Both carbon dioxide and water produced in the combustion reaction are greenhouse gases. That is, their presence in the atmosphere prevents heat from radiating back into outer space. Thus they are being implicated in global warming.

The physical state of a lubricant or fuel as it relates to temperature is an important property of that material. For example, if a material is to be used as a fuel in internal combustion engines, it should be a gas or liquid over the entire range of temperatures experienced on earth. As was indicated earlier, the length of the carbon chain in a molecule is a large factor in determining its melting point. The long carbon chain molecules that make up a portion of crude oil are essentially useless as fuels or lubricants because they are solids at room temperatures. The petroleum industry has a process called catalytic cracking whereby such molecules are broken into smaller molecules thus making them more useful. Because of catalytic cracking, the amount of gasoline that can be extracted from a barrel of crude oil has been greatly increased.

As indicated earlier in this chapter, the length of the carbon chain making up a molecule has a large effect on its physical properties. **The longer the chain, the higher the viscosity and melting or boiling point**. For example, methane (one carbon atom) boils at −164 °C (−295 °F), octane (eight carbon atom chain) boils at 125 °C (237 °F). Therefore, methane is a gas at room temperature and octane is a liquid at room temperature.

Plant oils have another factor that enters into the viscosity and melting point of the oil. That factor is the presence of one or more double bonds between carbon atoms making up the carbon atom chain of the oil. **Plant oils typically have one, two, or three double bonds between carbon atoms that make up the fatty acid chain of the oil**. In general, as the degree of unsaturation increases, the melting point decreases. This property of plant oils is important as we consider the gelling properties of lubricants. Unsaturated plant oils tend to remain in the liquid state at lower temperatures than do saturated hydrocarbons.

Applications like cooking and some forms of lubricants require solid fats rather than liquids. To remedy this situation, manufacturers use hydrogen to "saturate" unsaturated plant oils. Eliminating the double bonds in plant oils greatly increases their melting points thus converting them to solids at room temperature. Margarine is prepared from plant oils by a process called ***hydrogenation***. Basically, hydrogen gas in the presence of a catalyst is added to the double bonds in plant oils converting them to saturated fats. The effect is to create a solid out of a liquid

```
        H
        |
    H—C—O—H
        |
        H
```

Figure 2.9 Methanol or methyl alcohol

oil. This process increases the longevity of the oil in frying applications and could also be helpful in creating more stable vegetable oil-based lubricants.

If oxygen atoms are chemically bonded to hydrocarbon molecules, a whole new group of interesting chemical compounds results. Oxygen can form two chemical bonds. For example an oxygen atom inserted between a hydrogen atom and the carbon atom of methane results in alcohol. The structural formula for this molecule is shown in Figure 2.9.

The name of this compound begins with the parent hydrocarbon from which it is derived (methane), dropping the "e" ending and adding "ol." Thus, the name of the above compound is methanol. Note that the "meth" part of the name indicates one carbon atom. A commonly used alternate name for methanol is methyl alcohol.

Ethanol is the alcohol derived from ethane (Figure 2.10).

Ethanol, also referred to as ethyl alcohol, is prepared by a process called fermentation. Most of us are familiar with ethanol as being the beverage alcohol. Recently, ethanol has been produced by fermenting corn or other starchy grains added to gasoline to make a product called gasohol. The percent of ethanol in gasohol can vary up to 85%. As an example, the "E" in E-85 gasohol refers to the percentage of ethanol in the fuel as being 85%. Pure ethanol absorbs water from its surroundings and thus cannot be used "straight" as an engine fuel without special modification to the fuel handling equipment.

The presence of the –OH group generally makes the material more chemically active.

Hydrocarbons for example are not very reactive whereas alcohols are much more reactive chemically. Alcohols are therefore subject to the often detrimental reactions like oxidation.

When an alcohol comes in contact with oxygen it is oxidized to an organic acid. Vinegar (a three or five percent solution ethanoic acid or acetic acid) is formed when ethanol comes in contact with oxygen. In wine making, if oxygen is allowed to enter the wine-making vessel, the resulting liquid will become vinegar. Acids, like the vinegar from oxygenated wine, have a sour taste and the ethanol is oxidized to acetic acid. The structural formula for acetic or ethanoic acid is shown in Figure 2.11).

The functional group responsible for the acidity in the molecule is the –COOH group, that is, the group in which carbon is double bonded to oxygen and at the same time bonded to an –OH group. **When the COOH group occurs at the end of a long hydrocarbon chain,**

```
      H  H
      |  |
  H—C—C—O—H
      |  |
      H  H
```

Figure 2.10 Ethanol or ethyl alcohol

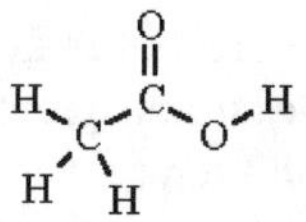

Figure 2.11 Structural formula for acetic or ethanoic acid

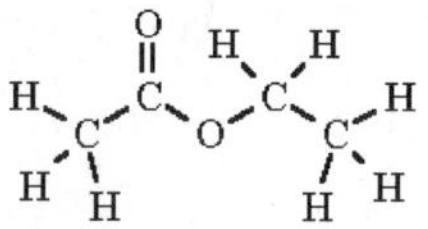

Figure 2.12 Ethyl acetate

the substance is referred to as a fatty acid. Fatty acids are components found mixed with the oils of oilseed plants and present problems in the formulation of biodiesel from these oils. Organic acids tend to be more corrosive than alcohols. Therefore, when oxidation occurs in a lubricant to form an organic acid, the resulting material is more detrimental to the device being lubricated.

Another class of organic compounds essential to the biolubricants area is a class of compounds called esters. **An ester is formed when an organic alcohol reacts with an organic acid**. Esters often have very pleasant aromas and are responsible for the "nose" in a glass of wine or the flavor of fruits and vegetables. Wintergreen, for example, is an ester derived from methanol and salicylic acid. If one were to react ethanol with acetic acid, an ester would form (Figure 2.12). This compound called ethyl acetate (ethyl from the alcohol, acetate from the acid) is used as a solvent to remove finger nail polish. HOH (water) is lost when an acid and an alcohol react together. This is an important observation. **Whenever an ester is formed from an acid and an alcohol, water is the byproduct of the reaction**. The entire reaction can be reversed in that water can be added to an ester to regenerate the original acid and alcohol.

One of the components of biodiesel fuel is methyl oleate. The material is an ester that is formed when methanol is reacted with oleic acid.

If the –OH group is removed from methanol and the –H is removed from the oxygen atom at the end of the oleic acid molecule and then joined together, water (HOH) is formed. The other substance formed in the same reaction is methyl oleate (Figure 2.13) where the $-CH_3$ is joined to the oxygen atom vacated by hydrogen to form an ester of methanol and oleic acid. As will be described later this kind of a chemical reaction is referred to as a "methathesis" reaction. This reaction is common to biodiesel formation. The fatty acid can be one of a number of such

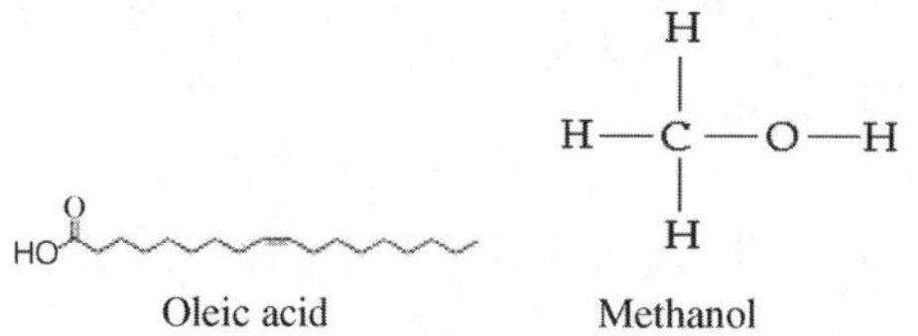

Oleic acid　　Methanol

Figure 2.13 Components of methyl oleate

substances derived from plant oils. The alcohol is generally methanol or ethanol, methanol being the most common for its lesser cost and greater ease of reactivity. You will read of FAMES (fatty acid methyl esters) in the literature.

These concepts apply to the chemistry of lubricants. Some of the above concepts are repackaged as practical pointers.

2.2.1 Pointers for Non-Chemists on Vegetable oil and General Chemistry

Below is a list of general points that are known to confuse students and chemistry knowledge seekers.

1. A single BOND is made of two electrons (not a single electron) one from each of the atoms bonded together. Thus a PAIR of electrons for a single bond.
2. CARBON is unique in the fact that it can BOND to itself, meaning two carbon atoms can each donate an electron to the bond (a pair of electrons) to form a single bond.
3. Single BONDS made up of TWO electrons, one from each atom, form the most common bonding in hydrocarbons or plant oils.
4. A single BOND, made up of TWO electrons, one from each atom, forms the most stable carbon to carbon bond.
5. A DOUBLE BOND between carbon atoms requires that each atom contribute two electrons to the bond resulting in four electrons (two pairs of electrons) being shared in total.
6. When four electrons, two from each atom are shared (double bonding), a point of instability arises. Double carbon to carbon bonds are much more unstable chemically than are single carbon to carbon bonds.
7. When six electrons, three from each carbon atom are shared, a triple bond forms resulting in even greater instability.
8. Fatty acids are made of carbon chains containing mostly single carbon to carbon bonds. Fatty acid chains vary in length from four carbon atoms to as many as 36 carbon atoms.
9. If the fatty acid has one double bond, *cis* or *trans* isomers are possible. (See later examples of *cis* and *trans*.)
10. Glycerol is a triahydric alcohol meaning that it contains three hydroxy groups.
11. Vegetable oils and fats can be in the form of monoglycerides, diglycerides, or more commonly triglycerides (Figure 2.14), depending on how many of the three hydroxy groups on glycerol have been replaced by fatty acids.

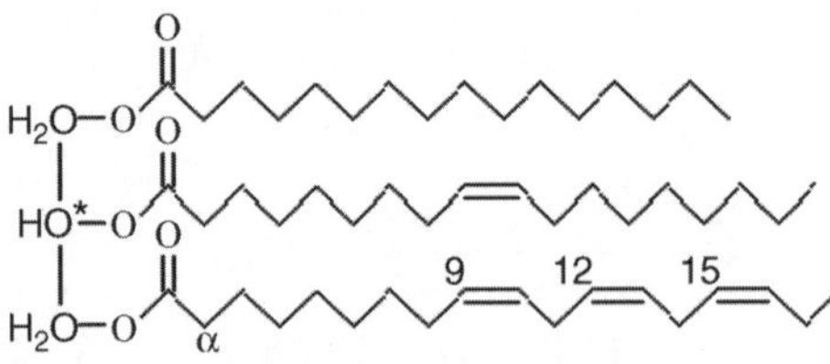

O
H_2O–O O
HO*–O O
9 12 15
H_2O–O α ω

Figure 2.14 A triglyceride showing positions of double bonds

Table 2.1 Saturated fatty acids [1]

Number of carbon atoms	Common name of acid	Systematic name	Melting point (°C)
C 4	butryic	butanoic	−8
C 6	capoic	hexanoic	−3.4
C 8	caprylic	octanoic	16.5
C 10	capric	decanoic	31.5
C 12	lauric	dodecanoic	43.5
C 14	myristic	tetradecanoic	54.4
C 16	palmitic	hexadecanoic	62.9
C 18	stearic	octadecanoic	69.6
C 20	arachidic	eicosanoic	75.4
C 22	behenic	docosanoic	79.9
C 24	lignoceric	tetracosanoic	84.2
C 26	ceratinic	hexacosanoic	87.7

12. Fatty acids may have several double bonds, but rarely more than four double bonds.
13. If a fatty acid has NO double bonds, them it is considered a SATURATED fatty acid. Table 2.1 shows saturated fatty acids.
14. If there is only ONE double bond in the fatty acid's carbon chain, then it's MONOUNSATURATED; and if more than one double bond exists in its chain, it is considered POLYUNSATURATED – examples are shown in Tables 2.2 and 2.3.

    In Table 2.4, percentages of monounsaturates, polyunsaturates and saturates for some common vegetable oils are presented.

    Table 2.5 shows various fatty acids and known sources of those fatty acids in known oils.
15. The melting point of triglycerides is important in formulating industrial lubricants. The variation of CHAIN LENGTH and the DEGREE OF UNSATURATION combine to form many triglycerides with different melting points (Figure 2.14).
16. Two UNSATURATED fatty acids may have the same number of double bonds and formulas, but due to the placement of their double bonds, they may have different chemical properties. For example, melting point is impacted as shown in Table 2.3 for oleic (13 °C), elaidic (44 °C), and vaccenic (39 °C).

Table 2.2 Polyunsaturated fatty acids

Number of carbon atoms	Common name of acid	Systematic name	Melting point (°C)
C 18	linoleic	*cis*-(9,12)-octadecadienoic	−5
C 18	linolenic	*cis*-(9,12,15)-octadecatrienoic	−11.3
C 20	arachidonic	*cis*-(5,8,11,14)-eicosatetraenoic	−49.5

Table 2.3 Monounsaturated fatty acids

Number of carbon atoms	Common name of acid	Systematic name	Melting point (°C)
C 10	caproleic	decenoic	—
C 12	lauroleic	dodecanoic	—
C 14	myristoleic	*cis*-9-tetradecenoic	−4
C 16	palmitoleic	*cis*-9-hexadecenoic	0–5
C 18	oleic	*cis*-9-octadecenoic	13
C 18	elaidic	*trans*-9-octadecenoic	44
C 18	vaccenic	*trans*-11-octadecenoic	39
C 22	erucic	*cis*-13-docosenoic	33

17. Only EVEN CARBON fatty acids with a *cis* conformation occur naturally. Fatty acids containing an odd number of carbon atoms rarely occur.
18. To describe fatty acids, the TOTAL NUMBER OF CARBON ATOMS is stated, like "C 18" for stearic and "C 16" for palmitic.
19. To describe UNSATURATED fatty acids, the total number of CARBON ATOMS are mentioned as well as the number of DOUBLE BONDS; for example, oleic is C 18:1 (meaning 18 carbons in the chain with one double bond mono-unsaturated); linoleic is C 18:2 (meaning 18 carbons in the chain with two double bonds), and linolenic is C 18:3 (meaning 18 carbons in the chain with three double bonds). As a reminder, the more double bonds the less [oxidatively] stable the fatty acid.
20. Fat refers to triglycerides containing free fatty acids, mono-, and di-glycerols, phospholipids, waxes, cholesterol, and vitamins. These are collectively referred to as LIPIDS. Triglycerides (fats) are made up of 95% or more fatty acids esterified to glycerol.
21. An aldehyde is an organic compound containing a terminal carbonyl group (a carbon to oxygen double bond). It consists of a carbon atom bonded to a hydrogen atom and double-bonded to an oxygen atom (chemical formula: CHO).

Table 2.4 Mono- and polyunsaturates and saturates in common vegetable oils [2]

Oil	Monounsaturates	Polyunsaturates	Saturates
Olive	75%	11%	14%
Canola	58%	36%	6%
Peanut	48%	34%	18%
Palm	39%	10%	51%
Corn	25%	62%	13%
Soybean	24%	61%	15%
Sunflower	20%	69%	11%
Cottonseed	19%	54%	27%
Safflower	13%	78%	9%
Coconut	6%	2%	92%

Table 2.5 Chemical names and descriptions of common fatty acids. Reproduced by permission of Antonio Zamora

Common name	Carbon atoms	Double bonds	Scientific name	Source
Butyric acid	4	0	butanoic acid	Butterfat
Caproic acid	6	0	hexanoic acid	Butterfat
Caprylic acid	8	0	octanoic acid	Coconut oil
Capric acid	10	0	decanoic acid	Coconut oil
Lauric acid	12	0	dodecanoic acid	Coconut oil
Myristic acid	14	0	tetradecanoic acid	Palm kernel oil
Palmitic acid	16	0	hexadecanoic acid	Palm oil
Palmitoleic acid	16	1	9-hexadecenoic acid	Animal fats
Stearic acid	18	0	octadecanoic acid	Animal fats
Oleic acid	18	1	9-octadecenoic acid	Olive oil
Ricinoleic acid	18	1	12-hydroxy-9-octadecenoic acid	Castor oil
Vaccenic acid	18	1	11-octadecenoic acid	Butterfat
Linoleic acid	18	2	9,12-octadecadienoic acid	Grape seed oil
Alpha-linolenic acid (ALA)	18	3	9,12,15-octadecatrienoic acid	Flaxseed (linseed) oil
Gamma-linolenic acid (GLA)	18	3	6,9,12-octadecatrienoic acid	Borage oil
Arachidic acid	20	0	eicosanoic acid	Peanut oil/fish oil
Gadoleic acid	20	1	9-eicosenoic acid	Fish oil
Arachidonic acid (AA)	20	4	5,8,11,14-eicosatetraenoic acid	Liver oil
EPA	20	5	5,8,11,14,17-eicosapentaenoic acid	Fish oil
Behenic acid	22	0	docosanoic acid	Rapeseed oil
Erucic acid	22	1	13-docosenoic acid	Rapeseed oil
DHA	22	6	4,7,10,13,16,19-docosahexaenoic acid	Fish oil
Lignoceric acid	24	0	tetracosanoic acid	Small amounts in fats

References

1. Magnusson, G., Hermansson, G., and Leissner, R. (eds) (1989) *Vegetable Oils and Fats*, Karlshamns Oils and Fats AB, Sweden by Halls Offset, Vaxjo, ISBN 91-7970-927-3.
2. Gapinski, R.E., Joesph, I.E.,and Leyzell, B.D. (1994). *A Vegetable Oil Based Tractor Lubricant*, SAE Technical Paper Series #941758, SAE Publications, Warrendale, PA.

3

Petroleum-based Lubricants

3.1 Introduction

The National Petrochemical Refiners Association surveys oil refiners annually, and has published detailed information on the number of gallons of various industrial and automotive lubricants produced in the Unites States. Based on a review of the data from NPRA, about 2.5 billion gallons of lubricants are produced in the United States annually with 1.1 billion gallons being *industrial* lubricants and another 1.4 billion gallons considered *automotive* lubricants. While this is a significant volume, it is only about 1% of the total petroleum used in the United States.

The majority of petroleum lubricant base oils are produced through the refining and modification of crude oil. Crude oil, extracted from the earth, is a very complex mixture of various chemicals ranging in complexity from simple gaseous molecules to very high molecular weight components. Asphaltic components, for example, are very heavy materials left over after many of the lighter hydrocarbons are extracted from crude oil.

Base oils are derived from light and heavy crude, based on the type and sources where the crude petroleum is extracted. The presence of bitumen (a coal-like material) and heavy distillates in the oil make them denser than other components and are what make the oil **heavy** [**crude**]. Some heavy crude could contain over 70% bitumen. These oils are suitable for heavier petroleum derivatives like base oils for lubricants. **Light** crude, on the other hand is refined for gasoline and other fuels. Crude oils from the Middle East are usually light crude; while crude oils from other parts of the world including continental Europe and Asia, North and South Americas, and the North Sea are heavy crude.

This ***heavy/light*** classification is based on the composition and density. In addition to this classification, base oils are also divided into ***paraffinic*** and ***naphthenic***. ***Paraffinic oils*** come from *light crude* and contain a higher percentage of light gases. ***Naphthenic oils***, on the other hand, come from *heavy crude*. Oil refineries are either focused on making specialty oils such as base oil for hydraulic oils from heavy crude (naphthenic) oils, or they are focused on fuel production from light crude (paraffinic) oils.

Biobased Lubricants and Greases: Technology and Products, First Edition. Lou A.T. Honary and Erwin Richter

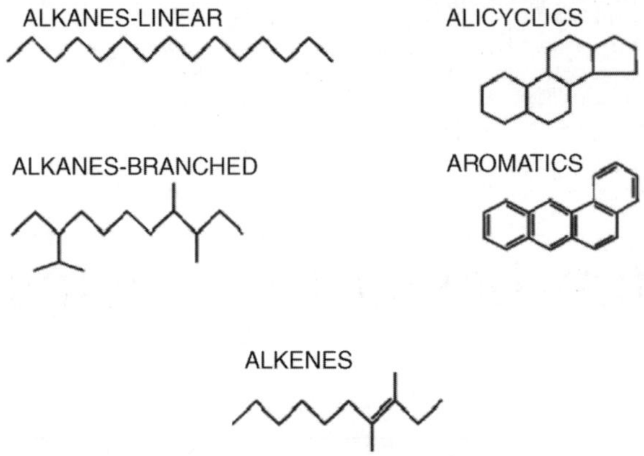

ALKANES-LINEAR
ALICYCLICS
ALKANES-BRANCHED
AROMATICS
ALKENES

Figure 3.1 Structure of hydrocarbons [2]

3.2 Basic Chemistry of Crude Oils

Like vegetable oils, mineral oils are primarily made of hydrocarbon, that is atoms of carbon and hydrogen bound together to form molecules of different structures. In addition to the paraffinic and naphthenic structures, these molecules could be aromatic or polycyclic aromatic. Figure 3.1 shows the basic structure of hydrocarbons as in the *Handbook of Naphthenic Specialty Oils for Greases Handbook* [1].

3.2.1 The Paraffinic Oils

Oils (alkanes) have either straight or branched structures (Figure 3.2). These oils contain waxes resulting in higher viscosity at low temperatures. During the process of refining, the normal alkanes or waxes are removed. This process is similar to winterization of vegetable oils, where the oil is cooled to the point where waxes are solidified and then filtered out. As some waxes may require much colder temperatures to solidify, some refineries are equipped to de-wax the oil by mixing the oil with a solvent. The oil is then cooled and allowed to crystallize, and the wax removed through filtration. These "deep de-waxing" processes can produce paraffinic oils with pour points down to −30 °C (−22 °F).

3.2.2 The Naphthenic Oils

Oils (alkenes) have cyclic structures and are sometimes referred to as cycloalkanes (Figure 3.2). They have often six-carbon rings (sometimes five- or seven-carbon rings) and offer great solvency properties as well as cold temperature flowability.

3.2.3 The Aromatic Oils

Oils have at least one ring of six carbon atoms with alternating double and single bonds (Figure 3.2). Most of the sulfur and nitrogen in the oil is bound to aromatic structures; giving the aromatic oils properties that are different from the straight chain paraffinic and the cyclic

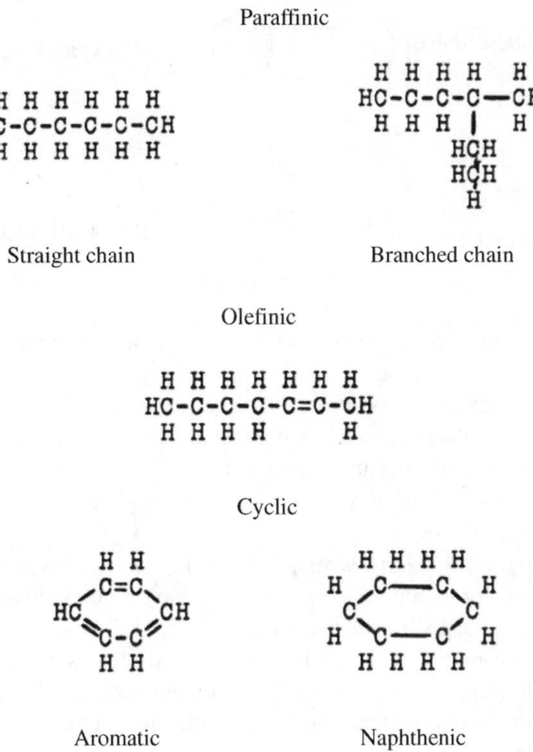

Figure 3.2 Representative hydrocarbon structures [3]

naphthenic oils. When the oil has several cyclic aromatic rings adjacent to each other, it is referred to as polycyclic, as opposed to monocyclic, containing one cyclic ring. Cyclic aromatics and normal alkanes occur in oil as single molecules, while others appear in various molecular structures. Most other elements such as nitrogen, sulfur and oxygen molecules in the oils are bound to the aromatic structures. When the aromatic structures are separated from the oil, these elements, too, are removed. Each of these elements in small quantities in the oil could have some sort of impact, both positive and negative. For example, nitrogen acts as an oxidation inhibitor and as passivator for copper. Sulfur compounds provide high extreme pressure and anti-wear properties, but their presence in the oil can also cause corrosion in copper materials. Sulfur also inhibits oxidation by destroying peroxides [1]. Oxygen dissolved in oils can accelerate oxidation, although the percentage of oxygen in fresh oil is very low.

The refining of crude oil involves, through diverse methods, separating the different types of hydrocarbons from the crude oil. Simple refining can be done by merely heating the crude oil. As different hydrocarbons have different boiling points, they evaporate at different temperatures, and can be condensed and extracted. Gases such as methane may evaporate at room temperature, whereas gasoline, diesel fuel, and higher viscosity hydrocarbons such as lubricant base oils evaporate at respectively higher temperatures. The challenge for a refining operation is to control the temperatures in a way that would allow only one type of hydrocarbon to evaporate at a time for a purer product.

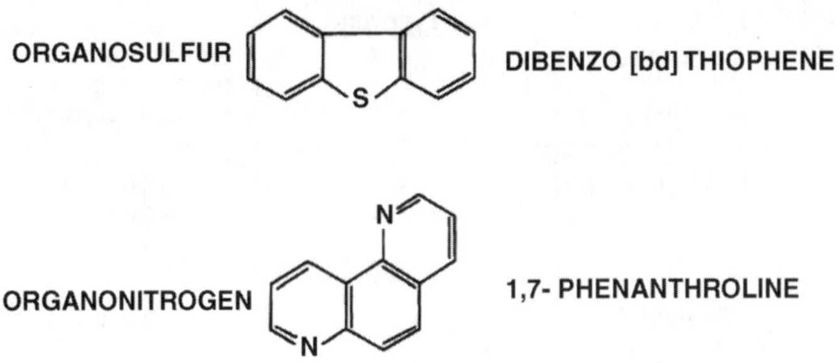

ORGANOSULFUR DIBENZO [bd] THIOPHENE

ORGANONITROGEN 1,7- PHENANTHROLINE

Figure 3.3 Examples of non-hydrocarbon molecules [2]

Crude oil mainly consists of saturated hydrocarbons. However, it also contains other materials such as alkenes, alicyclics, and aromatic compounds such as benzene or naphthalene.

Alkanes possess only carbon–carbon single bonds, which presents better resistance to oxidation and responds better to antioxidants as well as good viscosity/pressure properties compared to cyclic hydrocarbons.

Alicyclics (non-aromatic compounds) have a higher density and viscosity for their molecular weight when compared with alkenes. They have better cold temperature flowability and little or no wax to crystallize at cold temperatures. But, their viscosity pressure is not as good as the pure alkanes. Aromatics have the highest density and viscosity of the hydrocarbons. Aromatics also have good solvency property for additives but relatively poor oxidation stability.

Similarly, the non-hydrocarbon components in crude oil are typically hetrocyclic molecules that contain atoms other than hydrogen and carbon. These compounds are organosulfurs and organonitrogens with the organosulfurs being more prevalent than nitrogen- or oxygen-containing molecules. Examples of non-hydrocarbon structures are dibenzothiophene and 1,7-phenanthroline as organosulfur and organonitrogen respectively (Figure 3.3).

These hydrocarbon structures can be shown in both straight chain and in branched chains (Figure 3.2). As described in Chapter 1, carbon is somewhat unique among the elements in that it can form chemical bonds with itself. This property of carbon allows it to form carbon atom chains of various lengths. The carbons within the chain can combine with other elements, like hydrogen and oxygen. Of course, the length of the carbon chain is important in forming the structure of lubricants.

Typically, longer chains have a higher viscosity and shorter chains have a lower viscosity. Carbon chains of 25 carbons or more are in the range of base oils for lubricating oils. It should be noted that carbon chains of *lower* numbers have *higher* flammability. Thus gasoline (seven to nine carbon atom hydrocarbons) is a lighter, less viscous hydrocarbon and more flammable than diesel fuel (10–16 carbon atom hydrocarbons). Lubricating oil for example, $C_{25}H_{52}$ is heavier, more viscous, and less flammable than diesel fuel [3].

References

1. Nynas, A.B. (2009) *Naphthenic Specialty Oils for Greases Handbook*, Nynas Corporation, Sweden, www.Nynas.com.
2. Mortier, R.M.and Orszulik, S.T. (1992) *Chemistry & Technology of Lubricants*, Blackie and Sons Ltd., Glasgow.
3. National Lubricating Grease Institute (1996) *Lubricating Grease Guide*, 4th edn, National Lubricating Grease Institute, Kansas City, Missouri, USA.

4

Plant Oils

4.1 Chemistry of Vegetable Oils Relating to Lubricants

A major portion of oils and fats for human consumption come from plants and animals. The world production of edible oils is made up of 71% vegetable oils, 26% animal fat, and 2% fats from marine species. It is estimated that soybean oil, for example, contributes to nearly one-third of the world's oilseeds, with almost one-half produced in the United States.

According to the US Department of Agriculture, vegetable oils may be classified into three categories based on their production, use, and volume as (1) major oils, (2) minor oils, and (3) non-edible oils [1].

Major oils are those known for human or animal-feed consumption and often play important economic roles in the regions producing them. These include soybean, palm, rapeseed, sunflower, cottonseed, coconut, peanut, olive, palm kernel, corn, linseed, and sesame.

Minor oils are those known for their uses but do not match the large production magnitude of major oils. These oils have fatty acid profiles that could make them effective for industrial uses. They include niger, mango kernel, poppy, cocoa bean, shea, hempseed, grape seed, perilla, Chinese v. tallow, Ethiopian mahogany, German sesame, watermelon seed, avocado, and apricot seed.

Non-edible oils is the third group, and while the majority of oil plants are cultivated for food applications, a number of non-edible oils exist. These are linseed, castor, tung, and tall, and are commercially grown for their unique chemical makeup. These non-edible oils are used in industrial applications such as in soaps, paints, varnishes, resins, plastics, and agrochemicals. But, their use is also being considered for industrial lubricants applications. Examples of these oils include linseed, castor, neem, mahua, karanja, undi, kusum, khakan, pisa, kokum nahor, sal, and dhupa [1].

The popularity of biofuels over the past few years has resulted in a significant investment of public and private capital for the development of non-edible alternative crop oils. Although different in the end-use, many of the industrial crops and special processes developed for biofuels have applications in biobased lubricants as well. Additionally, the attention given to

Biobased Lubricants and Greases: Technology and Products, First Edition. Lou A.T. Honary and Erwin Richter

the negative health effects of trans fats has re-invigorated the development by major US seed companies of special varieties of oilseeds like the low linolenic and high oleic soybeans.

The Association for Advancements of Industrial Crops (AAIC) has a list of several alternative industrial crops that its members are working on. Most noteworthy are crops like *Cuphea*, *Camelina*, canola, castor, *Lesquerella*, peanut, and pennycress, among those being investigated by the oil crop division of AAIC. Some of these crops, like *Camelina*, have reached the commercial production stage and reasonably large acreages are being produced in the western United States. A brief description of such industrial crops will appear later in this chapter.

The future technologies will encompass the old fatty acids, the newer genetically enhanced high oleic varieties, and the more sophisticated, economically and chemically modified high functioning esters. These developments will not completely replace the use of petroleum for industrial and automotive lubricants, but they will capture a significant portion of those markets.

4.2 Triglycerides

The basic structure of fats and oils is an ester between glycerol and three fatty acids. Depending on the length of the carbon chain making up the fatty acid, the resulting fat can either be liquid (oil) or solid (fat or butter). Since glycerol is capable of bonding with three molecules of fatty acid, the result is referred to as a **triglyceride** (Figure 4.1).

A triglyceride, also called triacylglycerol (TAG), is a chemical compound formed from one molecule of glycerol and three fatty acids. The structure of glycerin is shown in Figure 4.2.

Glycerol is a trihydric alcohol (containing three -**OH** hydroxyl groups) that can combine with up to three fatty acids. Fatty acids may combine with any of the three hydroxyl groups to create a wide diversity of compounds (see Table 4.1). Monoglycerides, diglycerides, and triglycerides are classified as *esters*, which are compounds created by the reaction between (fatty) acids and alcohols (glycerol) that release water (**H_2O**) as a by-product.

Fatty acids contain a chain of carbon atoms combined with hydrogen (forming hydrocarbon). They terminate in a carboxyl group. If the three fatty acids are alike, the molecule is a *simple* triglyceride; if they are different it is a *mixed* triglyceride.

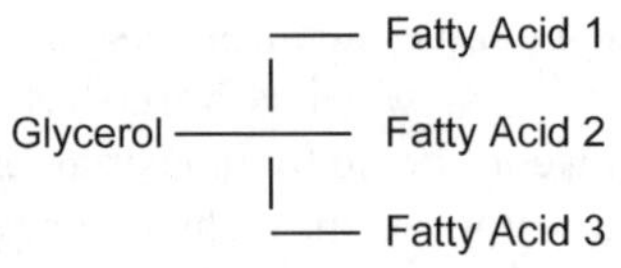

```
              —— Fatty Acid 1
             |
Glycerol ———————— Fatty Acid 2
             |
              —— Fatty Acid 3
```

Figure 4.1 Structure of a triglyceride

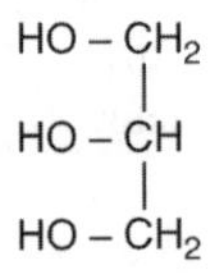

$$
\begin{array}{l}
HO - CH_2 \\
\quad\quad\ | \\
HO - CH \\
\quad\quad\ | \\
HO - CH_2
\end{array}
$$

Figure 4.2 Structure of glycerol or glycerin

Table 4.1 Fatty acid composition of selected fats and oils percent by weight of total fatty acids. Reproduced by permission of Antonio Zamora

Oil or fat	Unsat./Sat. fat ratio	Saturated					Mono-unsaturated	Poly-unsaturated	
		Capric acid C10:0	Lauric acid C12:0	Myristic acid C14:0	Palmitic acid C16:0	Stearic acid C18:0	oleic acid C18:1	linoleic acid (ω6) C18:2	Alpha linolenic acid (ω3) C18:3
Almond oil	9.7	—	—	—	7	2	69	17	—
Beef tallow	0.9	—	—	3	24	19	43	3	1
Butterfat (cow)	0.5	3	3	11	27	12	29	2	1
Butterfat (goat)	0.5	7	3	9	25	12	27	3	1
Butterfat (human)	1	2	5	8	25	8	35	9	1
Canola oil	15.7	—	—	—	4	2	62	22	10
Cocoa butter	0.6	—	—	—	25	38	32	3	—
Cod liver oil	2.9	—	—	8	17	—	22	5	—
Coconut oil	0.1	6	47	18	9	3	6	2	—
Corn oil (maize oil)	6.7	—	—	—	11	2	28	58	1
Cottonseed oil	2.8	—	—	1	22	3	19	54	1
Flaxseed oil	9	—	—	—	3	7	21	16	53
Grape seed oil	7.3	—	—	—	8	4	15	73	—
Illipe	0.6	—	—	—	17	45	35	1	—
Lard (pork fat)	1.2	—	—	2	26	14	44	10	—
Olive oil	4.6	—	—	—	13	3	71	10	1
Palm oil	1	—	—	1	45	4	40	10	—
Palm olein	1.3	—	—	1	37	4	46	11	—
Palm kernel oil	0.2	4	48	16	8	3	15	2	—

(continued)

Table 4.1 (*Continued*)

Oil or fat	Unsat./Sat. fat ratio	Saturated					Mono-unsaturated	Poly-unsaturated	
		Capric acid C10:0	Lauric acid C12:0	Myristic acid C14:0	Palmitic acid C16:0	Stearic acid C18:0	oleic acid C18:1	linoleic acid (ω6) C18:2	Alpha linolenic acid (ω3) C18:3
Peanut Oil	4	—	—	—	11	2	48	32	—
Safflower oil[a]	10.1	—	—	—	7	2	13	78	—
Sesame oil	6.6	—	—	—	9	4	41	45	—
Shea nut	1.1	—	1	—	4	39	44	5	—
Soybean oil	5.7	—	—	—	11	4	24	54	7
Sunflower oil	7.3	—	—	—	7	5	19	68	1
Walnut oil	5.3	—	—	—	11	5	28	51	5

http://www.scientificpsychic.com/fitness/fattyacids.html retrieved 18 April 2010.
Percentages may not add to 100% due to rounding and other constituents not listed.
Where percentages vary, average values are used.
[a]Non high oleic variety.

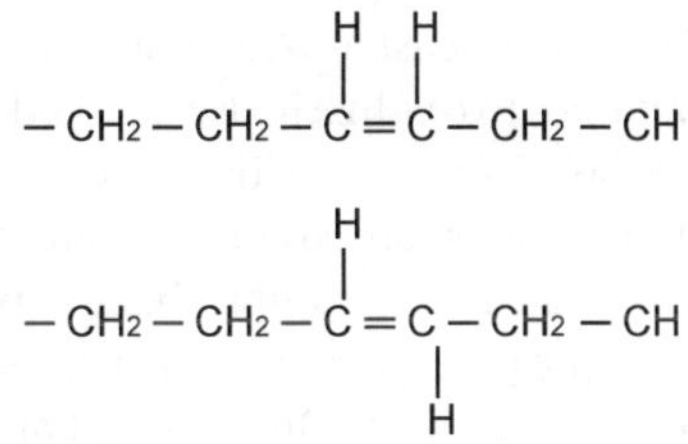

```
              H   H
              |   |
– CH2 – CH2 – C = C – CH2 – CH          cis

              H
              |
– CH2 – CH2 – C = C – CH2 – CH          trans
                  |
                  H
```

Figure 4.3 Hydrogen atoms on one side (*cis*) or on both sides of the chain (*trans*)

Saturated and unsaturated fatty acids. Each carbon atom along the chain has the ability to hold **two** hydrogen atoms. The fatty acid is **saturated** if there are no double or triple bonded carbon atoms in the chain. If a fatty acid chain contains multiple bonds it is **unsaturated**. In other words, by breaking the double or triple bonds, more hydrogen can be added to the molecule. The molecule is capable of holding more hydrogen atoms. An unsaturated carbon chain can be saturated by adding hydrogen to the multiple bonds in the chain. This process is referred to as **hydrogenation**.

Unsaturation is inversely related to the liquidity of the oil or its melting point and directly related to its solubility and chemical reactivity. With an increase in unsaturation, the melting point *decreases* (higher liquidity) while solubility in certain solvents and chemical reactivity *increases*. This usually results in oxidation and thermal polymerization. Saturated oils typically show more oxidative stability but have high melting points (lower liquidity). An example is palm oil, which has a high oxidation stability but is solid at room temperature and thus limiting its use for liquid lubricants applications unless modified.

For a normal, single-bond atom there is freedom of rotation around the bond, but there is rigidity at the site of a double bond. Thus two fixed positions of *cis* (meaning on same side) and *trans* (meaning across) are possible (Figure 4.3). *Trans* forms of fatty acids "pack" much closer together than do the *cis* forms. Therefore the *trans* forms more closely resemble the saturated fatty acids making "trans fat" more undesirable as a food oil. Because the saturated fatty acids have no double bonds to distort the chain, they pack more easily into crystal forms and, therefore, have higher melting points than unsaturated fatty acids of the same length. They are also less vulnerable to oxidation. This property also allows for winterization of the vegetable oils described later.

4.3 Properties of Vegetable Oils

Vegetable oils have many advantages and some shortcomings when considered for use in industrial lubricants and greases. Most importantly, unless modified, they lack oxidation stability. Oxidative stability of vegetable oils is dependent on the position and degree of unsaturation of the fatty acids that are attached to the glycerol molecule. For example, the majority of soybean oil fatty acid composition is comprised of conjugated carbon to carbon double bonds which make it more susceptible to oxidation. "Conjugated" is a term used to describe a condition where two double bonds in a carbon chain are close to each other. Conventional soybean oil contains approximately 52% linoleic acid, which has two conjugated double bonds, and 7%–8% linolenic acid which contains three conjugated double bonds. If left untreated the use of these oils could lead to increased oxidation and consequently increased viscosity. In extreme cases, if the oil continues to oxidize in use, it could lead to polymerization and formation of polymer films in the oil. To avoid oxidation in use, the vegetable oil is either chemically modified and/or antioxidants are used to increase oxidation stability. Hydrogenation, chemically adding hydrogen to the double bonds, is one method used to increase oxidative stability. Unfortunately, the melting point is also increased and can result in a product that is solid or semisolid at room temperature. Oilseeds that are genetically enhanced and have higher oxidation stability are more conducive for use in industrial lubricants and hydraulic oils.

The longer the fatty acid carbon chain, the higher the melting point. Double bonds within the carbon chain lower the melting point significantly. Vegetable oils, due to their fatty acid structure, tend to freeze at relatively higher temperatures than their mineral oil counterparts. A pour point comparison of hydraulic fluid using both mineral oil and soybean oil as base fluid is shown in Table 4.2. For applications where hydraulic oil or industrial lubricants are exposed to subzero temperatures, a mixture of vegetable oils and mineral or synthetic oils could be used. Mixing, however, impacts other properties of vegetable oils including viscosity index and flash/fire points as well as compatibility with elastomers and other components.

Vegetable oils, due to their polarity, adhere to metal surfaces for better metal-to-metal separation. Also, due to a higher viscosity index relative to petroleum oils, they are more stable as the temperature changes. For example, soybean oil has a viscosity index of about 220, with a viscosity of 30.69 at 40 °C (104 °F) and a viscosity of 7.589 at 100 °C (212 °F). Comparable

Table 4.2 Viscosity, viscosity index, and pour points of selected oils and identical hydraulic fluids utilizing soybean oil and mineral oil based fluids

Description	Pour point (°C) ASTM D 6749	Viscosity @ 40 °C ASTM D 445	Viscosity @ 100 °C ASTM D 445	Viscosity index ASTM D 2270
Refined high oleic soy oil	−16	31.19	8.424	200
Crude conventional soy	−6	31.69	7.589	222
Mineral oil- ISO VG 100	−50	20.58	3.684	28
Mineral oil- ISO VG 500	−32	96.21	9.040	53
Mineral oil blend of 57%-43% (of ISO VG 100 and 500	−49	37.95	5.295	53
Hydraulic fluid with crude conventional soy	−4	32.26	7.592	217
Hydraulic fluid with high oleic soy	−4	39.14	8.412	199
Hydraulic fluid with mineral oil blend	−11	25.24	4.248	46

naphthenic base oil with a viscosity of 37.95 at 40 °C and a viscosity of 5.295 at 100 °C would have a viscosity index of 53. Since the high viscosity index results in a more stable viscosity when temperatures change, lower viscosity vegetable oil-based hydraulic fluid could be used in applications where higher viscosity petroleum oil is required. As an example, ISO Viscosity Grade (VG) 46 hydraulic fluid made from vegetable oil may be suitable for applications where an ISO Viscosity Grade (VG) 68 from petroleum oil is specified. The authors' research has shown the difference between the viscosity of base soybean oils and formulated soybean oil-based and petroleum oil-based tractor hydraulic fluids (unpublished data). Table 4.2 shows the viscosity, viscosity index, and the pour points of soybean oils, mineral oils, and identical hydraulic fluid packages utilizing both soybean oil and mineral oil as base fluids. The mixture of ISO VG 100 and ISO VG 500 was prepared to create a viscosity range closer to soybean oils. The difference in viscosity index is significant with the soybean oil showing an almost 4 times higher viscosity index than that of petroleum mineral oils.

Tribological characteristics of the selected base oils and finished hydraulic fluids are shown in Table 4.3. Soybean oil shows better lubricating properties as indicated by the 4-ball wear (ASTM D 4172), 4-ball extreme pressure (ASTM D 2783), pin & vee, (ASTM D 3233A) and tapping torque (ASTM D 5619) results.

Table 4.3 Tribological characteristics of selected oils and finished hydraulic fluids

Description	Loadwear Index (weld point kg) ASTM D2783	4-Ball wear scar (mm) ASTM D4172	PIN and vee Force (lb) – Torque (lb-in) ASTM D3233 A	Tapping torque (N·m) ASTM D5619
High oleic soy oil	21.87 (160)	0.626	Broke @ 1755.64 lb-f Torque = 31.8	8.198
Crude conventional soy	26.74 (160)	0.589	Broke @ 1656.94 lb-f Torque = 53.1	8.027
Mineral oil ISO VG 100	12.63 (126)	0.663	Broke @ 567.64 lb-f Torque = 52.8	11.193
Mineral oil ISO VG 500	17.06 (126)	1.238	Broke @ 520.3 lb-f Torque = 93	11.073
Mineral oil blend of 57%:43% (of ISO VG 100 and 500	13.96 (126)	0.810	Broke @ 215.41 lb-f Torque = 51.1	10.99
Hydraulic fluid with crude conventional soy	33.98 (200)	0.510	N/A	N/A
Hydraulic fluid with high oleic soy	26.82 (160)	0.529	N/A	N/A
Hydraulic fluid with mineral oil blend	17.45 (126)	0.572	N/A	N/A

Table 4.4 Physiochemical characteristics of mineral oils

Properties	ISO VG32 Group I	ISO VG32 Group II	ISO VG32 Group III
Appearance	Clear yellowish liquid	Colorless liquid	Colorless liquid
Kinematic viscosity at 40 °C (cSt)	29.15	29.65	37.55
Kinematic viscosity at 100 °C (cSt)	5.14	5.37	6.43
Viscosity index	105	116	123

The flash and fire points of vegetable oils are consistently and considerably higher than equivalent viscosity mineral oils. Typically fire points of vegetable oils are greater than 300 °C (572 °F). This property is suitable for creation of industrial fluids and machinery greases that could meet some fire retardancy standards including those of the Factory Mutual standards in the United States. Metalworking fluids made from vegetable oils show less tendency to burn. Hydraulic applications like building elevators could benefit from the fire safety aspect of this property of vegetable oils. Table 4.4 presents a recent report on viscosity and the viscosity index of groups I, II, and III for comparison.

There are many factors that affect the fatty acid make-up of vegetable oils. In addition to their natural structure, changes in the growing conditions and geographic location and factors such as exposure to daylight, and light intensity and quality, impact the properties of vegetable oils. Because the fatty acid composition of oils and fats is unique, their characteristics are different. One important process that can be used to affect the types of fatty acid present is *partial hydrogenation* in which only some of the double bonds present in the carbon chain are given their full complement of hydrogen. Hydrogenation is explained later.

4.4 Vegetable Oil Processing

Since different vegetable oils can be processed nearly the same way, the processing of soybean oil which is the largest seed oil produced, is described here.

According to Mounts [2], extracting the oil from the oilseed requires three basic steps: (1) bean preparation, (2) oil extraction, and (3) solvent stripping and reclamation. After processing, the oil may be degummed for use in food-related applications. Small on-the-farm extruder/expeller units are finding popularity among farmers wishing to extract the oil out of the seed; with the residual meal used for feed purposes. These units do not use conventional oil-extraction techniques. Instead, the bean is forced through an extruder unit, which creates high pressure-induced temperatures. The shearing and grinding process results in the rupture of the cell walls of the bean. When the cell walls, particularly the oil cell walls rupture, they release some of the natural tocopherols, which have an antioxidant property, as well as some of the lecithin. Honary [3] reported that crude soybean oil obtained through this process showed more oxidation stability using ASTM D 7043 (formerly ASTM D 2271) hydraulic pump test than the crude oil obtained through conventional hexane-extracted processing. This could be attributed to the retention of the natural antioxidants (including tocopherols) in the oil by the extrusion process. The oil is then extracted from the meal by the use of an expeller, which is essentially a mechanical press.

4.4.1 Degumming

Gums are made up of phospholipids and non-triglycerides. Degumming involves the removal of phospholipids and other non-triglyceride materials. A byproduct of degumming is lecithin, which can be used as an emulsifying agent. Treating crude soybean oil with caustic soda neutralizes free fatty acids, hydrolyzes phosphatides, and removes some colored pigments and unsaponifiable matter. Gums maybe soluble in the oil or insoluble. During storage, the insoluble gums settle to the bottom of the storage tanks and natural degumming takes place. Soluble gums require the use of weak acids and more active methods to remove them.

4.4.2 Bleaching

Bleaching uses activated earth to absorb pigments, oxidation products, phosphatides, soaps, and trace metals. This is an important step where natural clays or earths (Fuller's earth), which are basically hydrated aluminum silicates, are used for bleaching; or changing the color of the oil to a neutral color. The process is simple: the neutralized oil is mixed with the appropriate amount of clay, heated to the bleaching temperature, and then filtered.

4.4.3 Refining

The crude oil contains considerable amounts of soluble and insoluble matter, including gums. Refining is a purifying treatment designed to remove free fatty acids, phosphatides and gums, coloring matter, insoluble matter, settlings, and other unsaponifiable materials from the triglycerides [4]. Accordingly, oils such as soybean oil are refined through chemically refining, caustic refining, and physical or steam refining. Refining removes undesirable constituents such as free fatty acids, coloring and insoluble matter, phosphatides, and is used as a purifying process. This prevents foaming, smoking, and cloudiness when the oil is heated. For most industrial lubricant applications, the use of refined oil is almost a necessity.

4.4.4 Deodorizing

Deodorization is a process of removing volatile substances and converting the oil into a bland-tasting, clear liquid [4]. Noticeable flavor and odor will have essentially disappeared when the free fatty acid content is lowered to 0.01–0.03% and the protein content is miniscule. The process of deodorizing involves heating, steam stripping, and cooling the oil before exposing it to the atmosphere. The high temperature is needed to remove triglycerides, by allowing the less volatile flavor and odor components present in the oil to evaporate. These substances must be volatilized to condense and, subsequently, be removed from the oil. Stripping steam is added to increase the rate of this process. The process removes the free fatty acids but the content cannot be reduced below 0.005% because of interaction with the stripping steam. Steam distillation may also result in the conversion of up to 25% of the linolenic acid present from the *cis* form to the *trans* form due to the elevated temperature.

4.4.5 Interesterification

The fatty acids can be rearranged or redistributed on the glycerol backbone, often accomplished by using catalysts at reasonably low temperatures. The oil is heated and mixed with the catalyst at about 90 °C. There also are enzymatic systems used for interesterification. This does not change the degree of saturation or isomeric state of the fatty acids, but can improve the functional properties of the oil.

4.5 Oxidation

When exposed to oxygen most materials will have some degree of reaction with it resulting in oxidation. Vegetable oils, for example, are used for frying applications; exposure to heat, moisture from food, light and air would cause them to oxidize rapidly. For frying applications, attempts are made to stabilize the oil through hydrogenation or to use more saturated oils like palm oil. One indication of oxidation is the onset of rancidity, which can be recognized by smell. In industrial applications, as indicated earlier, increased viscosity could be used to note the onset of oxidation.

The change in viscosity mentioned above is irreversible. This is different to when the oil thickens up due to exposure to cold temperature. In the latter case, heating the oil would reverse it back to its original viscosity. In the case of oxidation thickening, a change in molecular structure takes place resulting in the initiation of the polymerization process. Once initiated, larger molecules continue to form propagating the polymerization. This will then continue until terminated by intervention or by completion of the full polymerization.

Figure 4.4 (left) shows crude soybean oil that was exposed to air and ambient temperatures and, over a period of a few months, the oil is fully polymerized. On the right, the picture shows a tube with small holes that was used to blow and disperse air (oxygen) into the oil while heating it to 165 °C (237 °F). Under these circumstances the oil was oxidized over several hours, not several months. The same crude oil sealed in a bottle away from light would remain liquid for several years. Creating a barrier to oxygen is an effective way of reducing oxidation. The containers of some edible oils or beauty products that contain vegetable oils or animal fats may be topped off with nitrogen in a process called *nitrogen dosing*. The nitrogen, being heavier than air, replaces the air in the container thus preventing any oxidation. Figure 4.5 shows a poorly formulated product polymerized in its container.

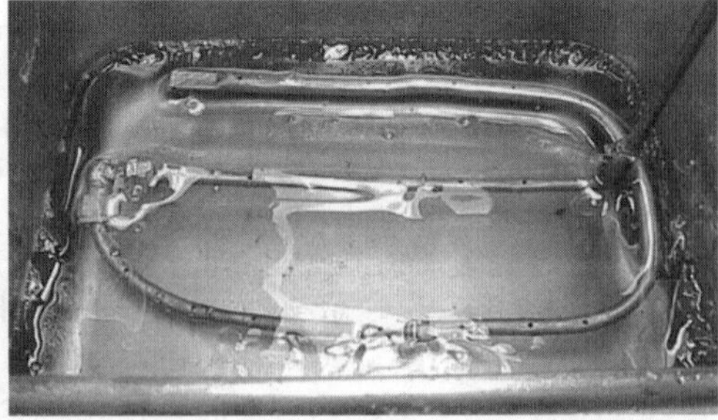

Figure 4.4 Soybean oil oxidized and polymerized naturally (left) or by using heat and air

Figure 4.5 Vegetable oil oxidized in storage, leakage on fitting, and on reservoir lid and walls

4.5.1 Reducing Oxidation

Oxidation of fats and oils lead to rancidity, and in industrial applications, could lead to increase viscosity and possibly to polymerization. The availability of double bonds and triple bonds in the oil will increase the possibility of oxidation. The oxidation rate for oleic acid, for example is 1/10 that of linoleic acid and 1/25 that of linolenic acid. Also, an increase in the temperature of oil increases the rate of oxidation. As a general rule **for every 10 °C rise in operating temperature, the oxidation rate doubles**.

Oxidation of fats results primarily in the formation of hydroperoxides, which are themselves tasteless and odorless, but decompose into volatile substances. When they decompose they produce aldehydes which have strong rancidity taste and odor. Oils containing oleic and linolenic acids form large number of aldehydes when they oxidize.

The food industry has several methods to prevent oxidation. These methods can be applied to industrial lubricants uses as well and include: (1) exclusion of air during processing, (2) cooling the deodorized oil before exposing it to the atmosphere, (3) preventing exposure of oil to air by a cover of nitrogen, and (4) addition of chemical antioxidants and metal excavengers [3]. Propyl gallate and tertiary butyl hydroquinone (TBHQ) may be added to increase oxidative stability, and *t*-butylhydroxy toluene (BHT) or *t*-butyl hydroxyanisole (BHA) may be added to increase shelf life. These additives may be used in amounts of up to 0.01% singly or 0.02% in combination. The oil is then filtered after the deodorization process has been completed to remove any solids that have formed or been introduced. After the process is completed, the free-fatty-acid content should be less than 0.03% (w/w) and peroxide values should be zero.

4.5.2 Hydrogenation

Since most triglycerides contain both saturated and unsaturated fatly acids, hydrogenation can be used to affect their fatty acid contents. Simply stated, **hydrogenation** is a way of saturating the double bonds. In the geometrical isomers, the *cis* structures are partially converted to the *trans* form. In the positional isomers, however, the original *cis*-9 double bonds, such as the oleic acid, are converted partially to a double bond at other positions in the chain [4,5].

Some oils like soybean oil have higher levels of **linolenic acid**, which is a polyunsaturated fatty acid. Its presence in the oil can lead to a higher degree of autooxidation, which, in turn results in off-flavor and odor (rancidity). When considering a vegetable oil for industrial lubricants, the oil with the lowest amount of linolenic and linoleic acids is more suitable. Through partial hydrogenation, the linolenic acid can be converted to oleic acids.

Direct addition of hydrogen to the double bond of an unsaturated fatty acid involves overcoming a considerable energy barrier. However, the introduction of a catalyst like rainy nickel greatly reduces the energy barrier. In the presence of a catalyst, the hydrogenation reaction proceeds at a much faster rate. Also, the removal of the reaction products from the surface of nickel requires overcoming a modest energy barrier before more energy is released. When hydrogenation takes place, the net energy release for a drop of one unit in iodine value is sufficient to raise the temperature of the oil by approximately 1.7 °C. This, of course, depends on the specific heat of the oil, which varies with temperature. The exothermic heat of reaction has been computed as 0.942 kcal per unit (1.7 BTU/lb). drop in iodine value [5].

The change in iodine value (IV) is used to express the degree of hydrogenation. The reason for this is because iodine reacts with double bonds. The greater the number of double bonds present, the greater the amount of iodine that will be consumed. The **iodine value** is defined as *the number of grams of iodine absorbed under standard conditions by 100 grams of fat*. It represents the degree of unsaturation in the fatty acid chain. The **saponification number**, defined as *the number of milligrams of potassium hydroxide required to saponify 1 g of fat,* is a measure of the average molecular weight of fatty materials. With partial hydrogenation, the amounts of linolenic and linoleic acid contents are reduced but the degree of saturation increases. The process of hydrogenation not only saturates naturally occurring *cis* double bonds but also isomerizes them to higher melting *trans* forms [6]. *Trans* fats used for cooking have been shown to have a negative impact on cholesterol levels and are thus considered unhealthy. A fully hydrogenated oil is solid at room temperature, like commercial brand Cresco. For industrial applications, the oil is normally *partially hydrogenated* to maintain the minimum desired liquidity while improving oxidation stability.

Partial hydrogenation can also increase in the amount of saturated fatty acid content in a given oil. By increasing the extent of hydrogenation, the oil begins to change physically and turn into semisolids and solids, which are suitable for margarine and shortening uses. Soybean salad oil, for example, which is partially hydrogenated to improve its shelf life, can be used for industrial uses because it still has good liquidity. However, where a lower pour point and higher degree of fluidity are desired, partially hydrogenated oil may need to be further "*winterized*" to reduce the amount of solids (high-melting triglycerides). The **pour point** is the *lowest temperature at which fluid movement can be detected.* The **cloud point** is the *temperature at which fats or other solids begin to crystallize when the liquid is chilled.* The cloud point is usually several degrees higher than the pour point. Both values are important at low temperatures. If the ambient temperature falls below the pour point, then the oil will not flow through the system.

The process of hydrogenation involves dispersing hydrogen gas into the oil as fine bubbles and using agitators to mix the oil and the gas. The oil is then cooled by coils containing heat transfer fluids that are used in heating and cooling of the mixture. When hydrogenation is complete, the oil is filtered until clear. Certain fatty acids will be hydrogenated more quickly, and each fatty acid has its own rate of reaction. Mono-, di-. and tri-unsaturated fatty acids undergo hydrogenation in sequential fashion. Varying the temperature, pressure, catalyst concentration, or agitation of the process can alter hydrogenation selectivity.

Temperature and hydrogenation – increasing the temperature decreases the selectivity of fatty acids hydrogenated, but it increases the formation of the *trans* form of fatty acids and increases the hydrogenation rate in general.

Pressure and hydrogenation – increasing the pressure decreases selectivity of fatty acids hydrogenated and decreasing formation of *trans* fatty acids, but it increases the rate of hydrogenation. In cooking oils, the presence of these *trans* fats which cause an increase in cholesterol levels, has become a serious health issue. As a result high oleic or low linolenic oils are finding popularity in cooking because they are stable enough that they do not require hydrogenation.

Agitation and hydrogenation – an increase in agitation decreases the selectivity of hydrogenation and decreases *trans* fatty acid formation, but it increases the rate of hydrogenation.

Catalyst and hydrogenation – an increase in the concentration of the catalyst leads to increases in selectivity of fatty acid hydrogenation, *trans* fatty acid formation, and the rate of hydrogenation.

Catalysts for hydrogenation – it is believed that flavor problems with vegetable oils are related to the linolenic acid content present. Possible remedies for this problem include using a copper catalyst during the hydrogenation process. This catalyst can produce an oil that has a linolenic acid content of less than 1%. Copper catalysts, however, are more sensitive to deactivating substances (such as free fatty acids, phosphatides, etc.) and are also less active than nickel catalysts.

Another option is to use a nickel-silver catalyst, which produces high linolenic selectivity during hydrogenation. Hydrogenation reactions are mostly accomplished using nickel catalysts. Copper catalysts are also used as an alternate to nickel, although copper catalysts are less active. By choice of conditions and the percentage of nickel catalyst, the unsaturated fatty acids begin a reaction sequence in which linolenic acid (18:3) is converted to linoleic (18:2), the linoleic acid is converted to oleic (18:1), and the oleic acid is converted to stearic (18:0) [2]. It should be noted that all of these conversions occur simultaneously, which means that during the conversion of linolenic acid to linoleic, the linoleic acid is also changing to oleic acid.

A manufacturer publication describing the use of various commercially available catalysts shows the effectiveness of a one nickel catalyst (G-135A) and its impact on hydrogenation and selectivity of fatty acids [7]. The study included the hydrogenation of soybean, canola, and fish oils. Using capillary gas liquid chromatography (GLC), the various fatty acids were measured and the IV calculated. The oil shown in Figure 4.6 is soybean oil with a maximum amount of unsaturated fatty acids content of 85.6%. As a result, breaking down the linoleic, linolenic or other acids could at best result in maximum conversion and retention to oleic acid of the same 85.6% at 75 IV. But, the process yielded maximum oleic acid content ranges from 67.6% to 76.7% based on the ranges of pressure and temperatures. This is almost 68–77% of the total unsaturated fats converted to oleic and stearic acids with the stearic acid content ranging from

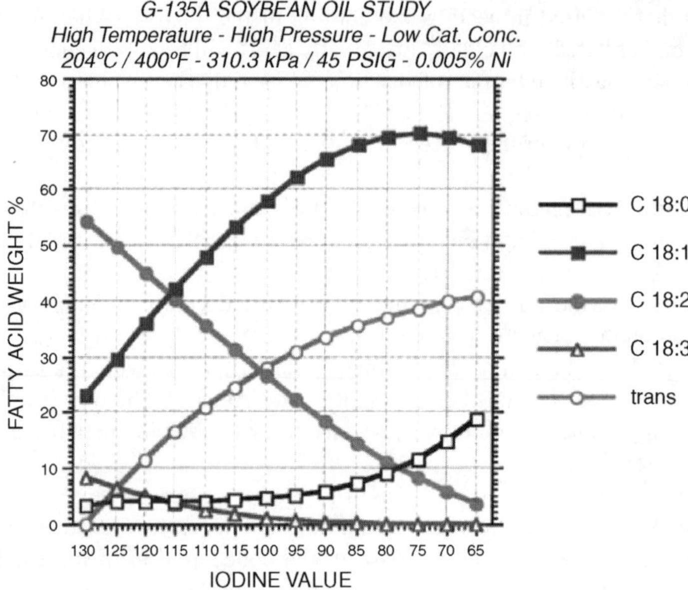

Figure 4.6 Formation of fatty acid in hydrogenation process. See Plate 2 for the color figure

nearly 9% to 13%. To determine the consistency and selectivity of the process, the ratio of the oleic to stearic acid can be monitored.

4.6 Winterization

Erickson *et al.* [4] provide a simple description of winterization, which involves chilling the fat at a prescribed rate and allowing the solid portions to crystallize. Then, through filtration the solids can be separated. If the oil is allowed to cool too rapidly, small crystals are formed that are more difficult to filter. Filtering cold oil, which presents a higher resistance to flow, requires energy and is time consuming, thus increasing the cost of the process. This process parallels dewaxing of petroleum oils, as explained earlier.

4.7 Chemical Refining

In the process of chemical refining, an alkali solution is used. The alkali combines with free fatty acids to form soaps that can be separated. This process is called saponification.

Phosphatides and gums absorb the alkali and are coagulated through hydration or degradation. Coloring is degraded, absorbed by gums, or made water-soluble by the alkali. The caustic process of refining involves analyzing the amount of free fatty acids and neutral oil in order to determine the amount of caustic soda to be added. The formation of different density layers occurs. One layer contains most of the oil and the other layer contains most of the unwanted particles. These layers are separated using a centrifuge. Physical refining does not use caustic soda and does not have soap stock.

Treatment with phosphoric acids removes phosphatides. This oil can then be used in steam refining. Steam refining uses a stripping steam to remove odor from the oil and also remove free fatty acids.

Another related refining process is known as the Zenith process. This process removes non-fatty-acid substances with concentrated phosphoric acid. The acid treatment removes calcium and magnesium from gums.

4.8 Conventional Crop Oils

Conventional crops are often commodity and domesticated crops that include extensive human intervention and improvement to create high yielding and nutritionally advanced products. An example is corn, which is grown in large quantities in the United States. Corn requires direct human intervention such as detasseling and pollination in order for crops to survive and for annual production to continue. Different regions of the world have domesticated their own indigenous crops and there are many crops that produce oil used for human consumption and could be useable for industrial uses.

A brief description of conventional vegetable oils now follows.

4.8.1 Soybean

Soybean (***Glycine max***) is the largest grown crop in the world and the United States is the leading producer of this crop (Figure 4.7). There are three species of soybean: ***Glycine ussuriensis*** (wild), ***Glycine max*** (cultivated), and ***Glycine gracilis*** (intermediate) [6]. It is the world's most important oilseed. In addition to the US, other countries producing soybeans include Brazil, Canada, Argentina, Paraguay, China, India, Indonesia, South Korea, Thailand, Italy, and Romania. In the United States, initially soybean was grown for its feed value for livestock. Soybean oil would have to be removed from the bean before it could be fed to the animals, although 2–5% oil left in the meal actually have added nutritional value as animal feed. Mechanical extraction of the oil often leaves some of the oil in the meal, whereas chemical extraction which includes stripping the oil using hexane can remove almost all of the oil from the meal. While in the past, soybean oil was often

Figure 4.7 Soybean field (left) and Soybeans ready for harvest. See Plate 3 for the color figure

considered a byproduct of meal, recently the value of the oil has also increased and some soybeans are grown for their oil.

The properties of commodity soybean oil include:

Density	0.9075
Viscosity	30.52 cSt @ 40 °C (104 °F) and 7.42 cSt at 100 °C (212 °F)
Viscosity index	224
Total acid number	0.10
Pour point	−9 °C (16 °F)
Flash point	314 °C (597 °F)
Gross heat of combustion (BTU/kg)	7623

Soybean constituents impacting oxidative stability were documented by Salunkhe *et al.* [1] and include more than 50% polyunsaturated fatty acids and about 15% saturated fatty acids, most of which is palmitic acid. The oil is usually hydrogenated to improve its oxidation stability; and sometimes is winterized if the resulting oil-based product is to be refrigerated. Optimized, hydrogenated soybean oils have been developed mainly for increased shelf life in the kitchen and stability during frying. The same technology can be applied for improving the oil's stability for industrial uses such as in metalworking fluids, hydraulic oils, and many other applications.

The main components in neutral lipids, phospholipids, and glycolipids are palmitic, linoleic, and linolenic acids. The average oil content on a moisture-free basis in soybean seed is about 20%. However, temperature has a marked effect on both polyunsaturated fatty acids and the oil content of soybean. Soybeans produced under controlled-temperature conditions showed oil contents of 23.2% at 30 °C (85 °F), 20.8% at −14 °C (7 °F), and 19.5% at lower temperatures. Basically, growing conditions, especially temperature, have a significant impact on the fatty acid profile of the beans. In the United States, most of the soybeans grown are processed into defatted meal flakes and crude oil. The meal is used mainly for animal feed and the oil is processed as edible vegetable oil. In producing biobased lubricants from soybean or other vegetable oils, unless the base oils are processed to produce a uniform quality, the year variations in the climate and growing condition must be taken into account. Formulations of the product may have to change from base oil to base oil if the origins of the oil and its growing conditions are not known. The importance of this variability in the industrial and machinery lubricants formulation cannot be overstated.

4.8.1.1 Changes Due to Processing

The *Handbook of Soy Oil Processing and Utilization*, which is a joint publication of the American Soybean Association and American Oil Chemists Society, provides extensive coverage of the storage and transportation requirements for soybean oils. Such information can be applied to most other vegetable oils as well. Processed vegetable oils, that are degummed, refined, bleached and deodorized, lose their natural anti-oxidants (like tocopherols) and may require the addition of antioxidants. Addition of 200 ppm TBHQ to vegetable oils would help reduce oxidation during transportation. For longer storage, processed oils are often topped with nitrogen to act as a barrier to air.

Figure 4.8 Palm, fruit. © 2007 ParaQuat Information Center, Syngenta Crop Protection AG and tree. Reproduced from Magnusson *et al.*, 1998 Vegetable Oils and Fats, Karlshamns AB, Sweden

4.8.2 Palm Oil

In addition to being a source of cooking oil, palm oil can be modified and used as fuel, lubricants, and other nonfood *new uses* such as palm ink. Oil palm (*Elaeis guilleensis* Jacq.) is an important, edible oil obtained from either palm fruit or from the palm kernel (Figure 4.8). Palm kernel is obtained as a minor product during the processing of oil palm fruit. Palm oil is solid at room temperature, has a neutral taste, and it can be extracted without the use of solvents thus meeting the standards for organic food processing. Honary [3] reported that untreated palm oil tested in ASTM D 2271 (currently ASTM D 7043) hydraulic pump test showed high oxidative stability for palm oil as measured by changes in the oil viscosity. Palm oil contains some triglyceride species that are completely saturated. It consists of mostly: monoglycerides (48–55%) and diunsaturated glycerides (30–43%) with small quantities of saturated (6–8%) and unsaturated glycerides (6–8%).

The fatty acid composition of palm *kernel* was reported by Godin and Spensley [10] to be:

Caprylic	3–4%
Capric	3–7%
Lauric	46–52%
Myristic	15–17%
Palmitic	6–9%
Stearic	1–3%
Oleic	13–19%
Linolenic	0.5–2.0%

These values change for palm oil produced in different parts of the world and various sources. Palm olein is the liquid fraction of palm oil and is used worldwide as cooking oil. The authors' research has shown the performance of an untreated palm olein as a hydraulic fluid with a supplier-reported fatty acid composition of lauric (0.2%), myristic (1.0%), palmitic (39.6%), stearic (4.6%), oleic (43.3%), linolenic (11%), and arachidic (0.3%). Accordingly, the oil contained 43.3%

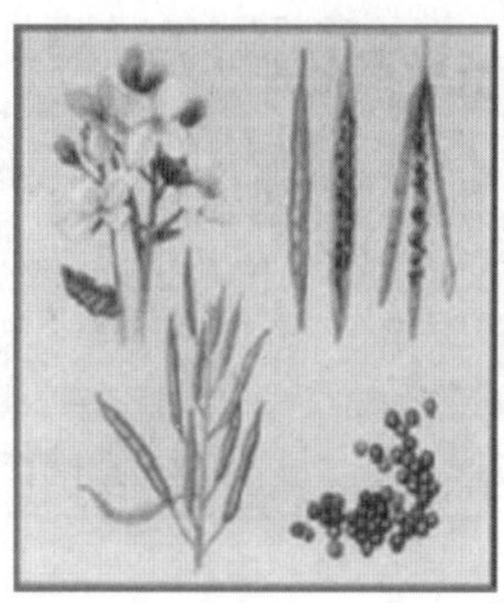

Figure 4.9 Rapeseed and canola. Reproduced from Magnusson *et al.*, 1998 Vegetable Oils and Fats, Karlshamns AB, Sweden

monounsaturates, 11.0% polyunsaturates and a high level of 45.7% saturates with 0% in *trans* acids. This type of oil showed excellent oxidation stability in a hydraulic pump test, but had the drawback of a very high melting point of about 23.9 °C (75 °F), being solid at room temperature.

4.8.3 Rapeseed

Rapeseed, typically grown in northern climates, thrives well in a cool, moist regions and is grown extensively in northern Europe and Canada (Figure 4.9). Canola oil, which is the low euricic acid version of rapeseed with near zero-erucic variety, contains 53% oleic acid and 11% linolenic acid. After soybean and palm, rapeseed is the largest produced vegetable oil in the world. Other countries growing rapeseed include India, China, Pakistan, and Australia. Like soybean, the meal obtained after the extraction of oil is used as animal feed. Rapeseed refers to more than one plant species and is often used to denote the seeds derived from oil-yielding members of the Brassica family, including some mustard seeds grown for edible or industrial oil (http://www.google.com/search?hl=en&sa=X&oi=spell&resnum=0&ct=result&cd=1&q=picture + of + rapeseed + plant&spell=1 retrieved 18 April 2010).. *Brassica napus* and *B. campestris* are the two most important and widely grown species. Summer types are grown in North America and a mixture of summer and winter types are grown in Europe. *B. juncea* and *B. campestris* are grown in India and the Far East. Wild populations of *B. campestris* have been reported from different regions of Europe and Asia. *B. napus*, is derived genetically from a natural hybridization of *B. campestris* and *B. oleracea* and occurs naturally in more restricted areas, mainly in Europe and North Africa.

Rapeseed contains about twice as much oil as soybeans, and the oil-free meal has only slightly less protein. Rapeseed oils exhibit a saponification value of 168–192 and an IV of 81–112. Others have shown that BHA/BHT with monoglyceride citrate to be ineffective against oxidation of canola oil. A polymeric antioxidant *anoxomer,* effectively inhibited oxidative change, in canola oil stored for 12 days at 65 °C when added at levels of 2000–4000 ppm.

4.8.3.1 Functional Properties of Rapeseed

Solubility. The isoelectric pH of rapeseed protein has been reported to be lower (pH 3.5) than soybean proteins.

Water absorption. The water-binding properties of a protein determine the extent of its interaction with water. Rapeseed protein products had a high water absorption.

Oil absorption. Canola meals prepared by the methanol-ammonia treatment have shown higher fat absorption than laboratory-produced hexane-extracted canola meals.

Approximate fatty acid composition of rapeseed (canola) oil:

Oleic acid	61%
Linoleic acid	21%
Stearic acid	2%
Alpha-linoleic acid	10%
Palmetic acid	4%
Other	2%

4.8.4 Sunflower Oil

Sunflower oil has a differing fatty acid composition if it is grown in northern climates as compared to southern climates. It is reported that oil from northern-grown seeds has a high linoleic acid content (64%) and low linolenic acid content (1%); whereas oil from southern-grown seeds contains a low linoleic acid content (49%) and a high oleic acid content (34% vs. 21%) [4].

There are two types of sunflower: oilseed and non-oil or confectionary (Figure 4.10). The confectionary type is consumed as whole roasted seed and represents less than 10% of total sunflower production [1]. The fatty acid composition of sunflower oil is primarily palmitic (7.2%), stearic (4.1%), oleic (16.2%), and a large portion of linoleic (72.5%) [11].

To extract the oil from sunflower seed, similar equipment and conditions to soybean oil extraction methods are used. After cleaning, drying, and dehulling, the oil is extracted by either mechanical extraction, prepress solvent-extraction, or by direct solvent-extraction

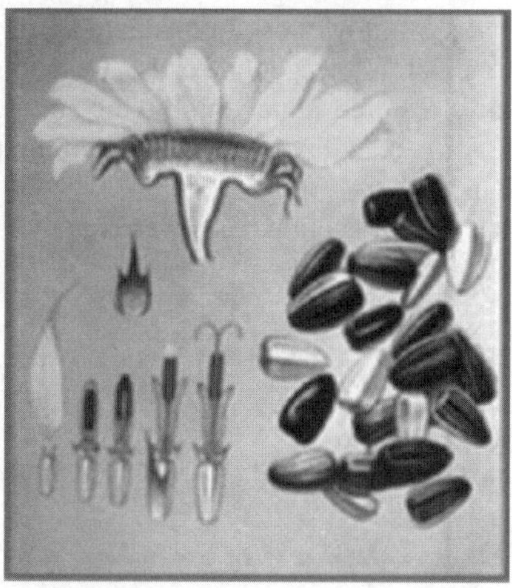

Figure 4.10 Sunflower and sunflower seed. Reproduced from Magnusson *et al.*, 1998 *Vegetable Oils and Fats*, Karlshamns AB, Sweden

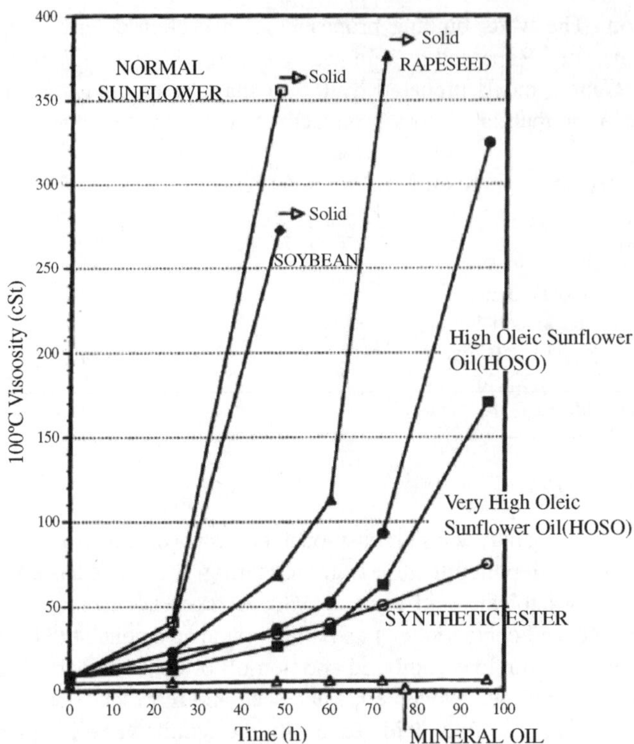

Figure 4.11 Viscosity increase in base vegetable oils in the presence of oxygen (at 10 l/h) at 100 °C (212 °F)

methods. Other processes such as bleaching, deodorizing, and winterization are also used to prepare the oil for food uses. Sunflower oil is considered premium oil (for food) as well as being one of the most palatable vegetable oils. Recent developments in genetic modification of the seeds have resulted in new high oleic and "ultrahigh oleic" sunflower oils with high oxidation stability. Naegley [12] reported on the performance of sunflower oil for industrial applications.

The physicochemical properties of sunflower oil were reported as follows [1]:

Density at 60 °C	0.89–0.90
Melting point (°C)	0
Smoke point (°C)	250[a]
Refractive index at 25 °C	1.4597
Iodine value	128
Saponification value	191
Free fatty acid (%)	0.01–0.03[a]
Unsaponifiables (%)	0.3–0.5
AOM time (hours)	10–15[a]

[a]For refined, bleached, deodorized oils.

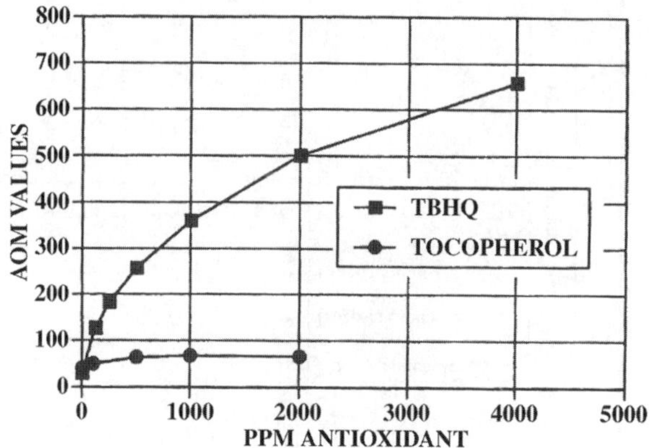

Figure 4.12 Effect of antioxidant TBHQ and tocopherol on the stability of high oleic sunflower oil

Naegley [12] compared the oxidative stability of several vegetable oils including high-oleic and very high-oleic sunflower oils in the presence of oxygen (Figure 4.11).

In order to further improve the oxidative stability of sunflower oil, Naegley compared the effects of two antioxidants tocopherol and TBHQ, on the oil using the active oxygen method (AOM) (Figure 4.12).

4.8.5 Corn

The oil content of the corn kernel (Figure 4.13), on a moisture-free basis, is about 5% as compared to soybean, which is about 20%. New genetic research has resulted in the development of the variety called high-oil corn, which contains about 8% oil. The high-oil corn can be directly utilized for commercial oil extraction. High-oil types are generally low

Figure 4.13 Corn. See Plate 4 for the color figure

Table 4.5 Fatty acid profiles of different vegetable oils

Typical fatty acid compositions of selected edible oils and fats

Fatty acid composition as a percentage of the fatty acid contents. For values >2 rounding-off has been calculated to the next half number. Fatty acids present only in trace amounts have not been included.

	Butyric	Caproic	Caprylic	Capric	Lauric	Myristic	Myristoleic	Pentadecanoic	Palmitic	Palmitoleic	Hexadecadienoic	Hexadecatrienoic	Hexadecatetraenoic	Margaric	Margaroleic	Stearic	Oleic	Linoleic	Linolenic	Octadecatetraenoic	Arachidic	Gadoleic	Eicosadienoic	Arachidonic	Elcosapentenoic (Epa)	Behenic	Erucic	Docosapentenoic	Docosahexaenoic	Lignoceric			
Carbon atoms; double bonds	4:0	6:0	8:0	10:0	12:0	14:0	14:1	15:0	16:0	16:1	16:2	16:3	16:4	17:0	17:1	18:0	18:1	18:2	18:3	18:4	20:0	20:1	20:2	20:4	20:5	22:0	22:1	22:5	22:6	24:0	Iodine value	Saponification	Carbon atoms; double bonds
Babassu oil			4.5	7.0	45.0	16.0			7.0							4.0	14.0	2.5													13–17	247–253	Babassu oil
Butter oil	3.2	2.0	1.0	2.5	3.0	11.0	1.0	2.0	27.0	2.0				0.5		12.0	28.5	3.0	0.5			0.5									25–42	236–252	Butter oil
Chicken fat					0.1	1.0	0.2	0.1	25.5	7.0				0.1	0.1	6.5	37.5	20.5	1.0		0.2	0.2									74–80	185–200	Chicken fat
Cocoa butter						0.1			26.0	0.4				0.2		35.0	34.0	3.0			1.0	0.1				0.2					34–40	190–200	Cocoa butter
Coconut oil		0.5	8.0	6.0	47.0	17.5			9.0							3.0	7.0	1.8	0.1		0.1										7–10	250–264	Coconut oil
Cotton seed oil					0.1	1.0			22.0	1.0				0.1	0.1	3.0	19.0	53.0	0.3		0.3					0.1					100–120	191–199	Cotton seed oil
Illipe fat									17.0							43.5	36.5	1.0			2.0										53–70	188–200	Illipe fat
Lard				0.1	0.1	1.5		0.1	26.0	3.3				0.4	0.2	13.5	44.0	9.5	0.4		0.2	0.6	0.1								48–65	195–202	Lard
Maize oil						0.1			11.0	0.1				0.1		1.4	28.0	58.0	0.9		0.3					0.1					111–131	188–198	Maize oil
Menheden oil						9.5	0.5	1.0	19.5	12.0	1.5	1.5	1.0	1.2		3.5	11.5	1.5	1.5	3.5	0.3	2.0		2.0	14.5		0.5	2.5	9.0		115–160	180–192	Menheden oil
Olive oil									10.0	0.5						3.0	77.0	8.0	0.5		0.5	0.5									79–90	185–196	Olive oil
Palm kernel oil (pko)		0.3	3.5	3.5	48.0	16.0			8.0							2.5	15.5	2.5			0.1	0.1									13–18	145–255	Palm kernel oil (PKO)

Palm oil				0.1	1.0			45.0	0.2	0.1		4.5	38.0	10.0	0.5	0.5			0.1			50–55	196–209	Palm oil
Palm oil-olein fraction				0.2	1.0			40.0	0.2			4.5	42.5	11.0	0.2	0.3			0.1			56 min.	196–209	Palm oil-olein fraction
Palm oil-stearine fraction				0.5	1.5			5.5				5.0	29.5	7.0	0.1	0.4			0.1			46 max.	196–209	Palm oil-stearine fraction
Peanut oil					0.1			10.0	0.2	0.1	0.1	3.0	42.0	38.0		1.5	1.0		3.0		1.0	84–95	184–195	Peanut oil
Pko-olein fraction	0.1	4.5	4.0	42.0	12.5			8.5				2.5	22.0	3.5		0.1	0.1					25–31	240–250	PKO-olein fraction
Pko-stearine fraction	0.1	2.5	3.0	55.0	20.0			8.0				3.5	7.0	0.8		0.1						6–9	240–250	PKO-stearine fraction
Rape seed oil (high erucic)					0.1			4.0	0.3			1.0	18.5	14.5	11.0	0.8	6.5	0.8	0.5	41.0	1.0	100–115	170–180	Rape seed oil (high erucic)
Rape seed oil (low erucic)					0.1			4.0	0.3	0.1		1.5	61.5	20.0	10.0	0.5	1.0		0.3	0.5	0.2	100–115	180–193	Rape seed oil (low erucic)
Safflower oil					0.1			7.0	0.1			2.5	13.0	76.5	0.3	0.2	0.1		0.2			126–152	172–195	Safflower oil
Safflower oil (high oleic)					0.1			3.5	0.1			5.0	81.5	7.5	0.1	0.5	0.2		1.2		0.3	82–92	172–195	Safflower oil (high oleic)
Sal fat								7.0				41.0	41.0	3.0	0.5	7.0			0.5			35–43	185–195	Sal fat
Shea fat								3.5				43.0	45.0	6.5		1.5	0.5					53–65	178–190	Shea fat
Soya bean oil					0.1			11.0	0.5	0.1		3.5	22.0	54.0	8.0	0.5			0.3			117–141	189–195	Soya bean oil
Sunflower seed oil					0.1			6.5	0.3	0.1		4.0	21.5	66.0	0.5	0.4	0.1		0.5			113–143	186–194	Sunflower seed oil
Sunflower seed oil (high oleic)								3.5	0.1			5.5	81.5	9.0		0.3			0.1			81–91	186–194	Sunflower seed oil (high oleic)
Tallow				0.1	3.0	1.0	0.5	24.5	4.0	1.5	0.8	18.5	42.5	2.5	0.7	0.2	0.3					40–55	282–286	Tallow

yielders and need to be genetically improved for better agronomical performance. The oil is concentrated in the germ and is recovered both by wet milling, in the production of starch, and by dry milling, in the production of grits, meal, and flour. More than 90% of the corn grain produced is processed and fed to animals in the Western world, whereas in Asia and Africa, almost all grain produced is utilized for human consumption by traditional processing without separating the germ [1].

Corn oil is a premium oil because of its high polyunsaturated fatty acid content and its low content (<1%) of linolenic acid. Leibovitz and Ruckenstein [13] list the fatty acid composition of corn oil as follows:

Lauric acid	0.1%
Myristic acid	0.2%
Palmitic acid	11.8%
Palmiloleic	Trace
Stearic acid	2.0%
Oleic acid	24.1%
Linoleic acid	61.9%
Linolenic acid	0.7%

The physicochemical properties of corn oil were reported [13] as:

Specific gravity	0.918–0.925
Density at 60 °C	0.892–0.897
Titer (0 °C)	18–20
Melting point (0 °C)	0
Refractive index at 25 °C	1.4596
Iodine value	103–133
Saponification value	187–195
Free fatty acids, as oleic (%)	0.03–4
Unsaponifiables (%)	1.2–2.8

4.8.6 Safflower

Safflower is grown in the southwest United States and many other parts of the world and has been considered for use in industrial applications. Safflower oil has high linoleic acid content (73%) and low linolenic acid content (1%), and therefore has a superior profile when compared to soybean or linseed oils when considering oxidation stability. According to Salunkhe *et al.* [1] the ranges for four important fatty acids of high oleic safflower oil were:

Palmitic acid	0.9–3.1%
Stearic acid	9.4–12.0%
Oleic acid	65.9–73.4%
Linoleic acid	11.5–23.82%

Many other vegetable oils have been explored for industrial lubricant and grease uses. Recently, due to environmental concerns, most farming groups in the United States have explored new uses of their vegetable oils. Table 4.5 [8] presents the fatty acid profiles of several vegetable oils including most of the aforementioned oils. These are products that are commercially available in the United States and are representative of the fatty acid compositions.

References

1. Salunkhe, J.K., Chavan, J.K., Adsule, R.N., and Kadem, S.S. (1992) *World Oilseeds: Chemistry, Technology, and Utilization*, Van Nostrand Reinhold, New York.
2. Mounts, T.L., and Kym (1985) in *Handbook of Soybean Oil Processing and Utilization*, (eds Erickson, D.R., Pryde, E.H., Brekke, O.L., Mounts, T.L., Falb, R.A), American Soybean Association, St. Louis, MO/AOCS Press, Champaign, IL.
3. Honary, L.A.T. (1995) Performance of Selected Vegetable Oils in ASTM Hydraulic Tests, SAE Technical Papers, Paper 952075.
4. Erickson, D.R., Pryde, E.H., Brekke, O.L., Mounts, T.L., and Falb, R.A. (1985) *Handbook of Soybean Oil Processing and Utilization*, American Soybean Association; St. Louis, MO/AOCS, Press, Champaign, IL.
5. Mounts, T.L. (1980) in Erickson, D.R., Pryde, E.H., Brekke, O.L., Mounts, T.L., and Falb, R.A. (Eds) *Handbook of Soy Oil Processing and Utilization*, 9th edn, American Soybean Association; St. Louis, MO/AOCS, Press, Champaign, IL.
6. Dutton, H.J. (1966) in Erickson, D.R., Pryde, E.H., Brekke, O.L., Mounts, T.L., and Falb, R.A.(1985) *Handbook of Soybean Oil Processing and Utilization*, American Soybean Association; St. Louis, MO/AOCS, Press, Champaign, IL.
7. United Catalyst Inc. (1994) *G-135A Hydrogenation Study: Soybean, Canola, and Dish Oils*, Company Published Manual, Germany.
8. Magnusson, G., Hermansson, G., and Leissner, R. (eds) (1998) *Vegetable Oils and Fats*, Karlshamns AB, Sweden.
9. Mutakas, D.K. (1980) Recovery of oil from soybeans, in *Handbook of Soy Oil Processing and Utilization*, American Soybean Association, Champaign, IL.
10. Godin, V.J. and Spensley, P.C. (1992) T.P.I. Crop Products Digest No.1, 1971, Tropical Products Institute, London, in *World Oilseeds: Chemistry, Technology, and Utilization* (eds J.K. Salunkhe, J.K. Chavan, R.N. Adsule, and S.S. Kadem), Van Nostrand Reinhold, New York.
11. Maiti, et al. (1988) Handbook of annual oilseed crops, in *World Oilseeds: Chemistry, Technology, and Utilization* (eds J.K. Salunkhe, J.K. Chavan, R.N. Adsule, and S.S. Kadem), Van Nostrand Reinhold, New York.
12. Nagley, P.C. (1992) Environmentally acceptable lubricants, in *Seed Oils for Future* (eds S.L. Mackenzie and D.C. Taylor) AOCS Press, Champaign, Illinois.
13. Leibovitz, Z. and Ruckenstein, C. (1983) Our experiences in processing maize (corn) germ oil. *Journal of American oil Chemist Society*, 395–399, in Salunkhe, J.K., Chavan, J.K., Adsule, R.N., Kadem, S.S. (1992) *World Oilseeds: Chemistry, Technology, and Utilization*, Van Nostrand Reinhold, New York.

5

Synthetic Based Lubricants: Petroleum-Derived and Vegetable Oil-Derived

5.1 Esters

Most natural oils, whether they are petroleum or vegetable based, contain impurities and waxes and thus their properties are not uniform. Synthetic products, on the other hand, are made when distinct and purer components of the oils are synthesized so that their properties are more uniform than in their natural forms. Esters are primarily defined as natural or synthetic. Natural esters are derived from vegetable oils or animal fats.

The fully synthetic ester base oils may be categorized in two groups: (1) diacid esters and (2) polyol esters, both of which are biodegradable with low toxicity. According to Randles *et al.* [1] many compounds have been investigated as base stocks for synthetic lubricants including: polyalphaolefins (PAO), alkylated aromatics, polybutenes, aliphatic diesters, polyesters, polyalkyleneglycols, and phosphate esters. Silicone, borate esters, perfluroethers and polyphenolene esters are in limited use due to their high costs and performance limitations.

Synthetic esters are made using acids and alcohols and due to their purity offer thermal stability and cold temperature performance far superior to their base materials. The chemistry of vegetable oils is generally similar in that they are triglycerides; or esters derived from glycerin and a range of fatty acids from C12 to C18 [2]. Figure 5.1 shows a triglyceride, a trimethylolpropane, and an adipate.

While synthetic esters are more expensive to produce, they provide the needed oxidation stability and cold temperature flowability in order to be considered for use in the more demanding lubricants applications.

Esters made from petroleum base stock are made using a fundamental process involving esterification, filtration, and distillation. Randles [1] describes the fundamental reaction for manufacturing esters as:

$$\text{acid} + \text{alcohol} \rightarrow \text{ester} + \text{water}$$

Biobased Lubricants and Greases: Technology and Products, First Edition. Lou A.T. Honary and Erwin Richter

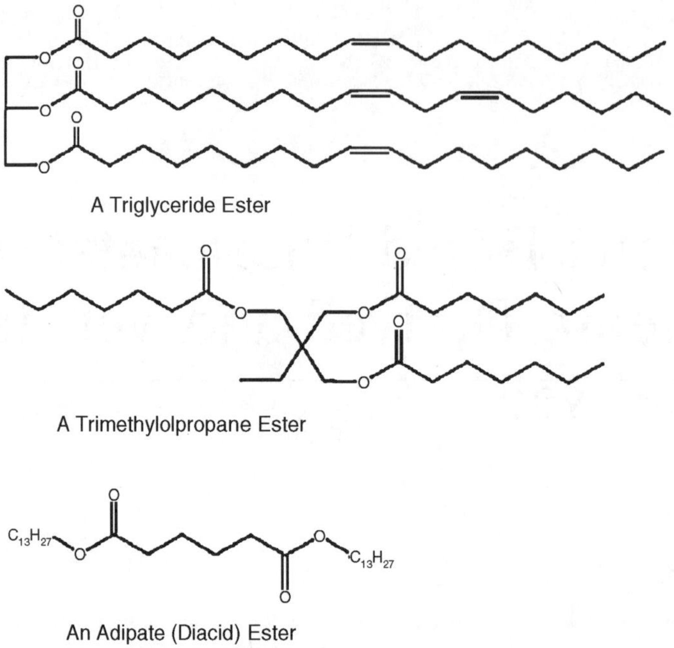

Figure 5.1 A triglyceride a trimethlolpropane, and adipate. Reproduced with permission of *Lubes-n-Greases* Magazine

with the reaction being reversible and driven to completion by the use of excess alcohol and removing water as it forms (p. 41). Accordingly, heat and catalysts are used to react the alcohol and acid. The possible catalysts used in this process can be sulfuric acid, *p*-toluene sulfonic acid, tetra-alkyl titanate, anhydrous sodium hydrogen sulfate, phosphorus oxides, and stannous octonate. To create a typical reaction, heat must be at 230 °C (446 °F) with 50–60 mmHg pressure. The alcohol and water vapors must be condensed and removed. At the conclusion of the reaction process, the untreated acid can be neutralized by sodium carbonate or calcium hydroxide and removed by filtration.

Similarly, reacting a polyhydric alcohol with a monobasic acid produces **polyol esters**. Examples of alcohols used include neopentylglycol (NPG), trimethylolpropane (TMP), or pentaerythritol (PE). Figure 5.2 shows some of the esters and their feedstock [3], and Table 5.1 lists the physical properties of these esters.

5.2 Esters for Biofuels

Another way to change the properties of a vegetable oil to become more comparable with those of low viscosity fuels and oils is by converting them chemically to monoesters. Diesel fuels have a viscosity of about 4 centistokes (cSt) at 40 °C (104 °F), and esters made for biofuels have similar level viscosities. Here too, the process used to make this conversion involves reacting an alcohol with the vegetable oil in the presence of a catalyst. In this process, from each triglyceride molecule, three mono-ester molecules and a glycerol molecule are obtained. The

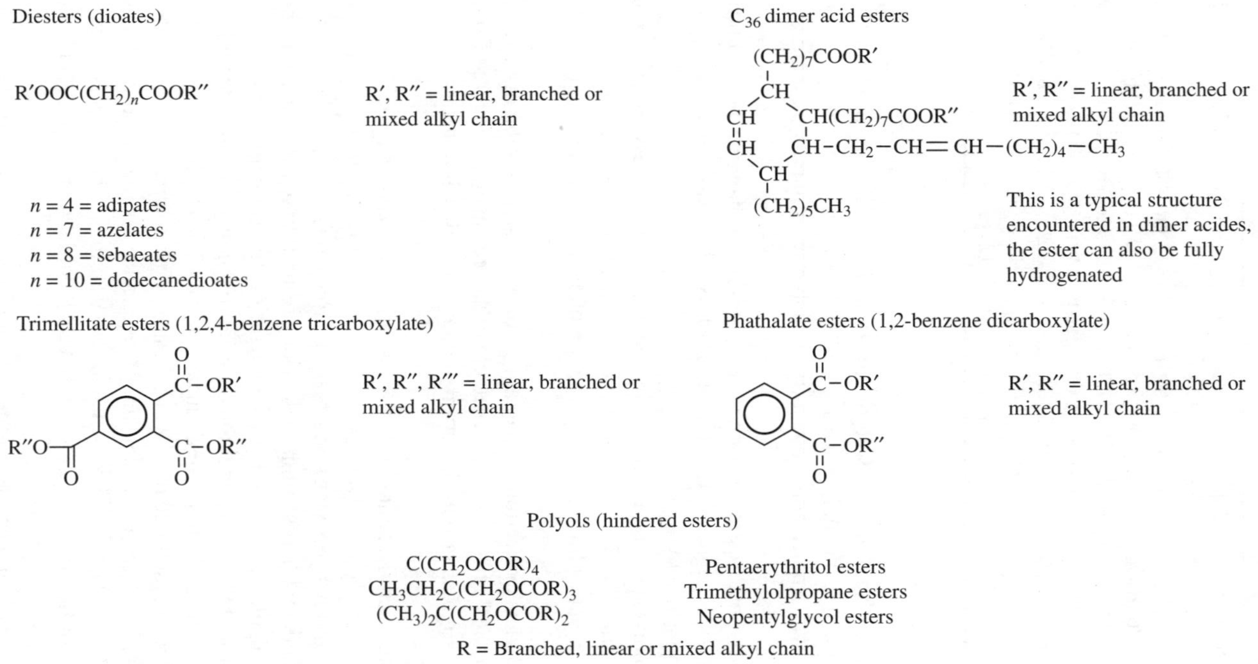

Diesters (dioates)

$R'OOC(CH_2)_nCOOR''$ — R′, R″ = linear, branched or mixed alkyl chain

$n = 4$ = adipates
$n = 7$ = azelates
$n = 8$ = sebaeates
$n = 10$ = dodecanedioates

C_{36} dimer acid esters

$(CH_2)_7COOR'$
CH
CH $CH(CH_2)_7COOR''$
CH $CH-CH_2-CH=CH-(CH_2)_4-CH_3$
CH
$(CH_2)_5CH_3$

R′, R″ = linear, branched or mixed alkyl chain

This is a typical structure encountered in dimer acides, the ester can also be fully hydrogenated

Trimellitate esters (1,2,4-benzene tricarboxylate)

R″O–C(=O)– ; C(=O)–OR′ ; C(=O)–OR″

R′, R″, R‴ = linear, branched or mixed alkyl chain

Phathalate esters (1,2-benzene dicarboxylate)

C(=O)–OR′ ; C(=O)–OR″

R′, R″ = linear, branched or mixed alkyl chain

Polyols (hindered esters)

$C(CH_2OCOR)_4$	Pentaerythritol esters
$CH_3CH_2C(CH_2OCOR)_3$	Trimethylolpropane esters
$(CH_3)_2C(CH_2OCOR)_2$	Neopentylglycol esters

R = Branched, linear or mixed alkyl chain

Figure 5.2 Selected esters and their feedstock

Table 5.1 Physical properties of esters [1]

	Diesters	Phthalates	Trimellitates	C36 dimer esters	Polyols	Polyoleates
Viscosity at 40 °C*	6 to 46	29 to 94	47 to 366	13 to 20	14 to 35	8 to 95
Viscosity at 100 °C*	2 to 8	4 to 9	7 to 22	90 to 185	3 to 6	10 to 15
Viscosity index	90 to 170	40 to 90	60 to 120	120 to 150	120 to 130	130 to 180
Pour point (°C)	−70 to −40	−50 to −30	−55 to −25	−50 to −15	−60 to −9	−40 to −5
Flash points (°C)	200 to 260	200 to 270	270 to 300	240 to 310	250 to 310	220 to 280
Thermal stability	Good	Very good	Very good	Very good	Excellent	Fair
Conradson carbon	0.01 to 0.06	0.01 to 0.03	0.01 to 0.40	0.20 to 0.70	0.01 to 0.10	?
% Biodegradability	75 to 100	46 to 88	0 to 69	18 to 78	90 to 100	80 to 100
Costs (PAO = 1)	0.9 to 2.5	0.5 to 1.0	1.5 to 2.0	1.2 to 2.8	2.0 to 2.5	0.6 to 1.5

*Viscosities in cSt

glycerol, a byproduct, is removed by water extraction. The final ester product is referred to as *methyl ester* if methyl alcohol is used or called *ethyl ester* if ethyl alcohol is used. These processes are well known and are readily available from numerous sources online and in print.

In organic chemistry, **transesterification** is the process of exchanging the organic group "R" of an ester with the organic group "R" of an alcohol. Acid or base catalysts can be added to make the process more reactive.

In the case of polyester, for example, in the synthesis process, a diester undergoes transesterification with diols for very large molecules. In this case, dimethyl teraphthalate and ethylene glycol react to form polyethylene teraphthalate and methanol, and this can be boiled off.

Biodiesel or methyl esterification is an example of transesterification of vegetable oils. It is easier to make biodiesel from unused vegetable oils because oils recovered from fast food restaurants or other sources require filtration and contain other impurities that would need to be removed.

It was stated earlier, that the simplest methyl esterification uses a vegetable oil, an alcohol, and a strong base as a catalyst. An example of this combination includes soybean oil, methyl alcohol, and lye (sodium hydroxide) and the reaction would occur at a slightly elevated temperature. The methanol and sodium hydroxide are first mixed together forming sodium methoxide. By mixing the sodium methoxide with the oil and allowing them to react, glycerin and the methyl ester will form. The addition of water to the reaction mixture will dissolve the glycerin and catalyst. The aqueous portion can be removed, leaving the methyl ester (biodiesel) behind. To have a biodiesel suitable for use in a diesel engine, the process needs to be thorough and the final products require significant quality assurance testing based on the established ASTM standards (ASTM D 6751). Problems observed in improperly prepared methyl esters include the presence of free fatty acids, which freeze at colder temperatures and plug the fuel line filter.

When using cooking oil or other used vegetable oils in addition to filtration, other steps can be taken to clean the oil and produce a cleaner methyl ester. The presence of free fatty acids in the oil during transesterification results in the formation of soap. When the lye is introduced during base-catalyzed trans esterification, the sodium ions combine with free fatty acids and form soap. In the presence of wash water, because the soap bonds with the methyl ester, the bonded ester washes off with the soap. While this can still be remedied, such a case would make the process cumbersome and thus inefficient.

In a modified process for used oil, first, the oil is esterified using an acid and then transesterified using an alcohol. The catalyst here is sulfuric acid and the alcohol is still the same methanol. In this phase of the reaction, lye is added to the mixture of the catalyst and methanol. During this phase, the sulfate ion in the sulfuric acid combines with the sodium ion to form sodium sulfate, which is a water-soluble salt and is removed in the wash. All the sulfur is then washed away from the methyl ester.

5.3 Complex Esters

Complex esters have a higher viscosity and a higher molecular weight than the common esters, offering advantages for some applications. Meng and Dresel (2001) explained the process for complex esters as first esterifying the diol with dicarboxylic acid, and then, depending on the desired product, reacting this ester with either carboxylic acid or a monoalcohol (p. 73) [4].

The *Journal of Synthetic Lubrication* has published numerous papers on synthetic esters. Some of these complex esters, for example, may be made via the reaction of a polyol, dicarboxylic acid, and monoalcohol as an end-capping agent. Key structural features of these esters and basic structure of an alcoxy group from the end-capping monoalcohols are presented in Figure 5.3 [5].

5.4 Estolides

The chemical modification and esterification of oils, especially of vegetable oils, is carried out to change the properties of the base material for special uses. High oxidation stability and lower cold temperature flowability are conditions most desired for vegetable oils in lubricant and grease uses. In an attempt to synthesize base oils that are suitable for use in formulating biobased lubricants, attempts are made to create flexibility in the use of the base raw materials. Estolides are designed with the idea of being able to use various vegetable oils or animal fats as starting materials. A patent by Steve Cermak *et al.* (2001), describes the creation of estolides for industrial and automotive lubricants.

Figure 5.4 shows the reaction scheme for oleic estolides as presented by Cermak and Isbell [6]. Estolides are formed "when the hydroxy fatty acid functionality of one fatty acid links to the site of unsaturation of another fatty acid to form oligomeric esters". The estolide number is accordingly defined as the average number of fatty acids added to the base fatty acid (Figure 5.5).

Figures 5.6 and 5.7 show the reaction scheme for complex estolides and 2-ethylhexyl esters as presented by Cermak and Isbell where oleic acid and various acids, butyric through stearic were treated with perchloric acid to produce a new class of saturated estolides with superior low temperature properties.

These estolides showed pour points of −30 °C (−22 °F) for the unsaturated oleic estolides and −40 °C (−40 °F) for the saturated ones, and oxidative breakdown of 200 and 400 minutes in the Rotary Bomb Oxidation Test (RBOT) compared with 200 minutes for comparable formulated mineral engine oil [6].

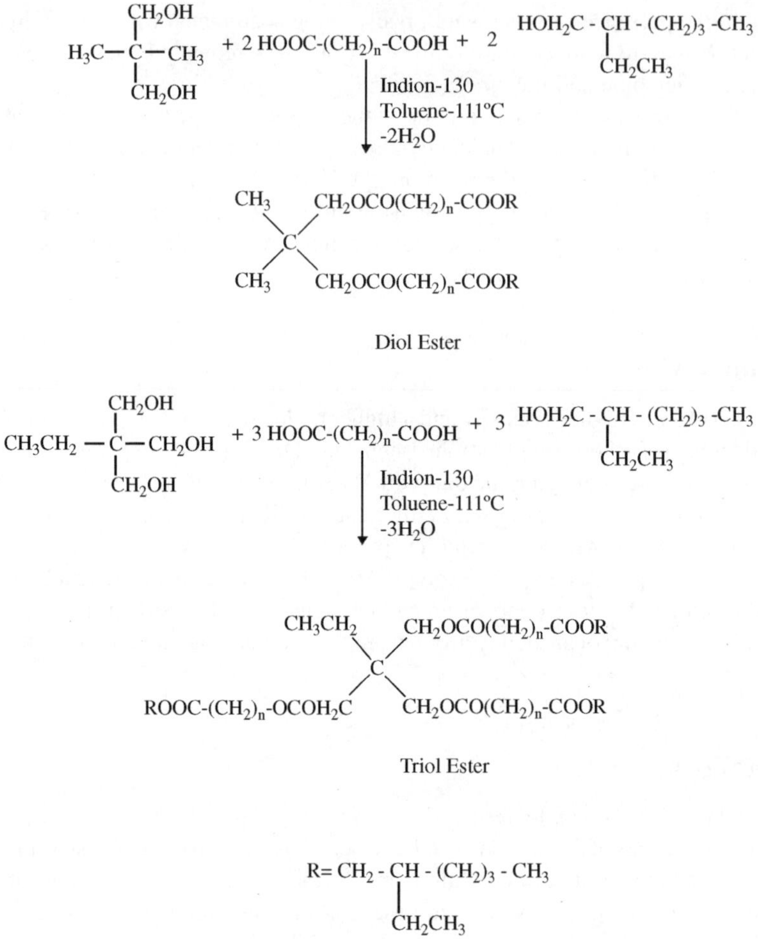

CH_2OH | $H_3C-C-CH_3$ | CH_2OH + 2 $HOOC$-$(CH_2)_n$-$COOH$ + 2 HOH_2C - CH(CH_2CH_3) - $(CH_2)_3$ -CH_3

Indion-130
Toluene-111°C
-2H_2O

CH_3, CH_3 > C < $CH_2OCO(CH_2)_n$-$COOR$, $CH_2OCO(CH_2)_n$-$COOR$

Diol Ester

CH_2OH | CH_3CH_2 —C—CH_2OH | CH_2OH + 3 $HOOC$-$(CH_2)_n$-$COOH$ + 3 HOH_2C - CH(CH_2CH_3) - $(CH_2)_3$ -CH_3

Indion-130
Toluene-111°C
-3H_2O

CH_3CH_2, $ROOC$-$(CH_2)_n$-$OCOH_2C$ > C < $CH_2OCO(CH_2)_n$-$COOR$, $CH_2OCO(CH_2)_n$-$COOR$

Triol Ester

R= CH_2 - CH(CH_2CH_3) - $(CH_2)_3$ - CH_3

Figure 5.3 Key structural features of these esters and basic structure of alcoxy group from the end-capping monoalcohols

Table 5.2 shows the oxidation stability and viscosity related data for three commercially supplied estolides. Distilled oleic estolide 2-ethylhexyl ester, oleic estolide 2-ethylhexyl ester and monomer, and distilled coco-oleic estolide 2-ethylhexyl ester.

Cermak and Isbell further provided comparative data on commercial petroleum based hydraulic oils, regular soybean oil, and estolides. The estolides without property enhancing additives show comparable viscosity index and lower cloud and pour points formulated products (Table 5.3).

Estolides are one example of chemical modification schemes that provide flexibility in the raw input materials and base oils with high oxidation stability and low pour point. It is anticipated that many of the future biobased lubricants and greases will be based on base oils made of complex esters, estolides and the like. These schemes are necessary for the larger use of biobased lubricants because they eliminate the property variation associated with growing

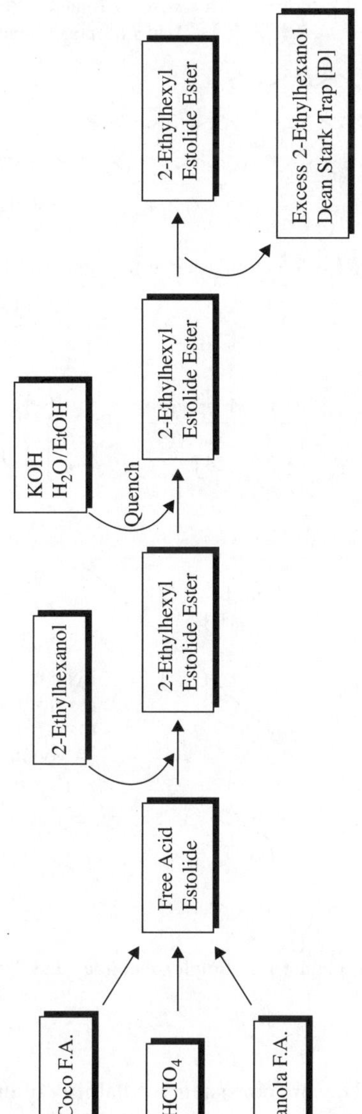

Figure 5.4 An example of a process for synthesizing estolides

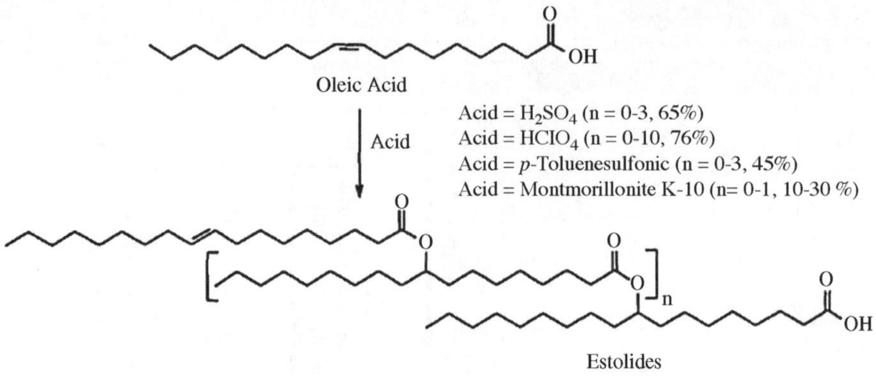

Figure 5.5 Reaction scheme for oleic estolide

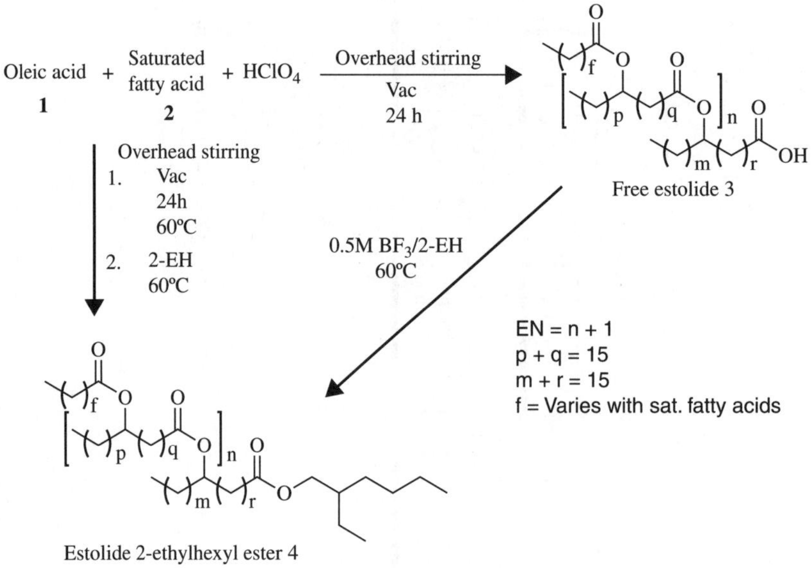

Figure 5.6 Reaction scheme for complex estolides and 2-ethylhexyl esters

conditions from year to year and ensure consistent availability of uniform quality and reliable base oils.

5.5 Other Chemical Modifications

There are many other processes being explored to modify the properties of fats or fatty acids and add value for industrial application. An example of such a process is **metathesis**.

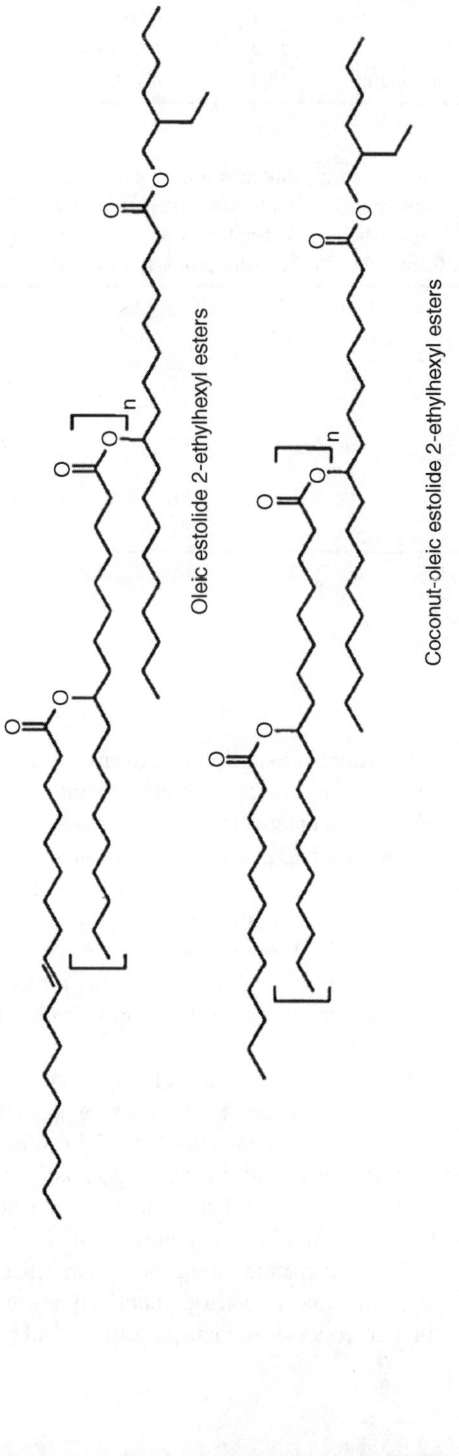

Oleic estolide 2-ethylhexyl esters

Coconut-oleic estolide 2-ethylhexyl esters

Figure 5.7 Reaction of oleic estolide 2-ethylhexyl ester and coconut-oleic estolide 2-ethlhexyl ester

Table 5.2 Oxidation stability and viscosity data of three estolides

ID #	Estolide type	OSI time	Viscosity at 40 °C	Viscosity at 100 °C	Viscosity index
1	Oleic estolide	18.18	98.17 cSt	15.40 cSt	167
2	Coco-oleic estolide	22.25	76.58 cSt	12.59 cSt	165
3	Oleic estolide with monomer	15.45	40.30 cSt	8.20 cSt	185

Table 5.3 Comparison of low temperature properties and viscosity index of coconut-oleic estolide 2-ethylhexyl esters to that of commercial lubricants. Reprinted from *Journal of Industrial Crops and Products*. Vol 18. Cermak, S.C. and Isbel, T. A. Synthesis and physical properties of estolide-based functional fluids, pp. 183–196. Copyright (2003) with permission from Elsevier

Lubricant[a]	Pour point (°C)	Cloud point (°C)	Viscosity @ 40 (°C)/(cSt)	Viscosity index
Commercial petroleum oil	−27	2	66	152
Commercial synthetic oil	−21	−10	60.5	174
Commercial soy-based oil	−18	1	49.6	220
Commercial hydraulic fluid	−33	1	56.6	146
Coco-oleic estolide[b]	−33	−26	55.2	162
Oleic estolide[b]	−33	−33	92.8	170

[a]Commercially formulated product.
[b]Unformulated estolides.

5.5.1 Metathesis

A simple definition of **metathesis** means to change places. In an example when two electrolytes are mixed together, the ions exchange places or exchange partners, so to speak. When two solutions are mixed, positive ions of one electrolyte encounter anions of the other. If this new pairing forms a more stable substance such as a solid or neutral molecule, the reaction is said to go to completion. Products that would lead to completion in a metathesis include the formation of a solid, the formation of water, or the formation of a gas.

The use of metathesis for the synthesis of organic compounds is considered a groundbreaking development. There are a large number of different organic molecules that are yet to be investigated for uses in new pharmaceuticals, base oil materials, and coatings, and are of high value and independent of petroleum.

The Nobel Prize in Chemistry for 2005 was shared among three scientists, Yves Chauvin of France, and Robert H. Grubbs and Richard R. Schrock of the United States, "for the development of the metathesis method in organic synthesis." Results of the work from the Laureates have already made significant impact in the chemicals industry, opening up new opportunities for synthesizing molecules that will streamline the development and industrial production of pharmaceuticals, plastics, and other materials. With their contributions, production through metathesis can become cheaper and more environmentally friendly [7].

In the practical application of metathesis, various catalysts are employed resulting in catalytic metathesis. The key in this process is to identify reliable and effective catalysts. The

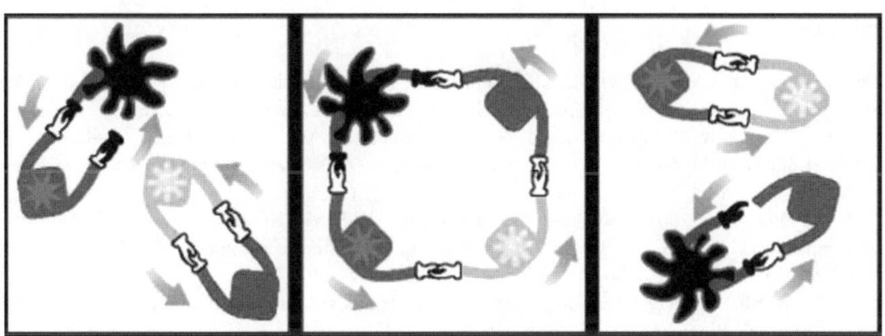

Figure 5.8 Metathesis concept illustrated as a change in dancing partners

researchers at a US based company known as Elevance Renewable Sciences provide Chauvins' mechanism in which the "catalyst pair" and the "alkene pair" dance round and change partners with one another a simple explanation of the practical use of catalytic metathesis. The illustration in Figure 5.8 describes this as a "dance," by which the "catalyst pair" and the "alkene pair" dance around and change partners with one another. The metal and its partner "hold hands" with both hands and when they meet the "alkene pair" (a dancing pair consisting of two alkylides) the two pairs unite in a ring dance. After a while, they let go of each other's hands, leave their old partners and dance on to their new partners. The new "catalyst pair" is now ready to catch another dancing "alkene pair" for a new ring dance or, in other words, to continue acting as a catalyst in metathesis.

More detailed information on catalytic metathesis is beyond the scope of this book. Its importance in creating new base oils for industrial lubricants needs to be emphasized.

5.5.2 Enzymatic Hydrolysis of Fatty Acids

The use of enzymes to mimic the breakdown of fats and triglycerides in the digestive system has been reported. Triglycerides can be enzymatically hydrolyzed to fatty acids and glycerol by the use of lipases. Most industrial hydrolyses involve high-pressure steam stripping to hydrolyze the triglyceride esters. Unfortunately, this process destroys some useful fatty acids found in the more exotic plant oils. The use of lipases for enzymatic hydrolysis of oils can provide a more efficient approach that is less energy intensive and does not alter the fatty acids that occur.

A US Patent (#5 089 403) by Hammond and Lee [8], describes using moistened, dehulled oat seeds or oat caryopses that contain lipases. When the oil is exposed to the moistened caryopsis, the fatty acids dissolve in the oil phase and the glycerol is absorbed in to the moist de-hulled oil seeds. According to Hammond and Lee, in a single cycle, about 20% of the oil can be catalyzed. Increasing the number of contact cycles will result in the hydrolysis of more oil.

Using lipases for splitting the triglycerides into free fatty acids and glycerin will have the potential for creating better lubricants and greases. The removal of glycerin from the fatty acids could reduce the hydrophilic properties of the formulated product. Also, if the free fatty acids are further processed into individual free fatty acids, more uniform grease soaps could be manufactured with predictable performance.

References

1. Randles, S.J., Stroud, P.M., Mortier, R.T., Orszulik, S.T., Hoyes T.J., and Brown M. (1992) in *Chemistry & Technology of Lubricants* (eds R.M. Mortier and S.T. Orzulik), VCH Publishers, inc., New York.
2. Bergstra, R. (2004) (Nov 2004) Green Means Go. *Lubes-n-Greases*, vol. 10, issue 11, pp. 36–42.
3. Mortier, R.M. and Orszulik, S.T. (eds) *Chemistry & Technology of Lubricants*, Blackie, USA and Canada and VCH Publishers, Inc. New York, p. 42.
4. Meng, T. and Dresel, W. (2001) *Lubricants and Lubrication*, Wiley-VCHGmbH, Weinheim.
5. Nagendramma, P. and Kaul, S. (2008) Study of synthetic complex esters as automotive gear lubricants. *Journal of Synthetic Lubrication*, **25**, 131–136.
6. Cermak, S.C. and Isbel, T.A. (2003) Synthesis and physical properties of estolide-based functional fluids. *Journal of Industrial Crops and Products* (18), 183–196.
7. Calderon, E.A., Ofstead, J.P., Ward, W.A.J., and Scott, K.W. (1968) *Journal of the American Chemistry Society*, **90**, 4133; Mol, J.C., Moulijn, J.A. and Boelhouwer, C. (1968) *Chemistry Communications*, 633; Calderon, N. (1972) *Acc. Chem. Res.* **5**, 127; Vetenskapsakademien – The Royal Swedish Academy of Sciences. (2005) "Development of the metathesis method in organic synthesis". Information Department, Box 50005, SE-104 05 Stockholm, Sweden. www.kva.seFrom http://www.elevance.com/ourtechnology.htm retreived April 18, 2010.
8. Hammond, E.G. and Lee, I. (1992) US Patent #5089403. Process for enzymatic hydrolysis of fatty acid triglycerides with oat caryopses.

6

Genetic Modification and Industrial Crops

6.1 Introduction

In addition to the chemical modification of oils which will play an important role in the future of biobased lubricants, there are other technologies that are showing promise. The genetic modification of crops, for example, was initially used primarily as a means of creating healthier oils that do not require hydrogenation for stability in cooking. The genetic modification of crops has taken a new direction with new and diverse goals. Genetic modification is considered for creating crops that are resistant to pests, herbicides, or aphids; or for including vitamins and pharmaceutical properties as well as transgenic properties for drought resistance or for season or climate independence. Interestingly, agronomists and food scientists, working on the genetics of oilseeds for the development of healthier food, may have contributed to the increased potential use of oilseeds in industrial lubricant applications. It is now clear that through genetic modification of the seeds, the fatty acid profile of the oilseeds can be altered with great benefits in terms of stability, cold temperature flowability and reduced need for chemical additives. This concept eliminates the need for hydrogenation or other chemical modification and thus reduces the cost of the base oil and finished products for use in industrial applications.

Genetic modification has been accomplished for some of the vegetable oils, including rapeseed, sunflower, and soybean. Salunkhe *et al.* [1] reported on the changes in the genetic makeup of rapeseed oil starting in the late 1960s. Accordingly, prior to the 1970s, rapeseed oil contained 20–50% erucic acid. The first Canadian low-erucic-acid variety, containing about only 3% erucic acid, was licensed in 1968. In 1974, a lower acid variety containing less than 0.3% erucic acid was introduced, following the Canadian government's encouragement to switch to low-acid varieties. CANada Oil Low Acids (CANOLA) is the Canadian version of the rapeseed with distinctly low erucic acid and low glucosinolate. It has a linoleic to linolenic acid ratio of 2 : 1. In the United States, various genetically modified seeds for canola, sunflower, and soybean have been developed by major seed and chemical

Biobased Lubricants and Greases: Technology and Products, First Edition. Lou A.T. Honary and Erwin Richter

companies. Currently, the availability of these oils with high oleic acids of 80–90% and excellent oxidation stability promise to create many new opportunities for vegetable-based industrial lubricants.

Perhaps the most promising new research advance to date, for biobased lubricants, is the development of new mutant lines of soybeans, which lead to new lines of high oleic soybeans. These mutant lines with *improved fatty acid profiles* of the oil can be cloned and then integrated into high-yielding elite lines of soybean seeds. Clone genes are introduced into soybeans to create transgenic lines with increased lysine, oleic, or stearic acid contents. These are oils with a very low content of polyunsaturated fatty acids. They show high oxidative stability, do not require hydrogenation for frying applications thus eliminating the formation of *trans* fats for healthier cooking oils.

Kinney [2] reported on the design of transgene constructs, which were assisted by using soybean somatic embryos in suspension culture as a model system for soybean seed transformation. The system has allowed the selection of those genes and promoters that are the most effective way of achieving the desired phenotypes in soybeans. According to Kinney, in soybeans, gene-transgene suppression is a more effective means of silencing endogenous genes than antisense. Sense suppression of genes encoding microsomal, fatty acid omega-6 desaturates, has resulted in soybean lines with over 80% oleic acid in their seed oil. This is over four times the oleic acid content of most commodity soybean oils. The high-oleic trait is reported to show stability in over at least three generations [2]. Accordingly, in one of the new strains, the oleic acid content was 81.3% and the seeds had an average of 7.6% palmitic acid, 5.9% stearic acid, 1.3% linoleic acid, and 3.1% linolenic acid. Honary [3] tested commodity soybean oil and a high oleic soybean oil in hydraulic pump tests using the ASTM D 7043 (formerly ASTM D 2271), which exposes the oil to 1000 hours of pumping pressure at 6894.76 kPa (1000 psi) and 79 °C (174 °F) with a fixed flow. As the oil is exposed to heat, air, moisture from the atmosphere, and metallic components it begins to oxidize as is observed by an increase in its viscosity. In earlier hydraulic pump tests, the author showed that changes in viscosity in the conventional soybean oil as being about 150% from 0 to 1000 hours while the high oleic soybean oil having almost 1.3% change in the same period. The same two oils tested in the oxidative stability instrument showed an oil stability index (OSI) of 7 hours for the commodity oils and 192 hours for the high oleic soybean oils (Figure 6.1).

6.2 Industrial Crops

The increased demand in crop-based oils for use in fuel and lubricants has created the opportunity to explore the potential use of crops that are grown naturally in various parts of the world and are not necessarily edible. The research into non-edible secondary crops has further intensified due to the food vs. fuel controversy. As petroleum prices increased to over $140 per barrel in 2008, the increase in food prices were attributed to the growth of biofuels during the first half of the decade. As a result, considerable efforts were made in identifying and qualifying natural oil crops with oils suitable for industrial and fuel uses. More importantly, when such crops are grown in arid and otherwise unusable lands, they offer an attractive alternative to both edible oils and petroleum.

In the United States, the Association for the Advancement of Industrial Crops (AAIC) has been leading the exploration of the use of non-edible native crops that have oil and the potential

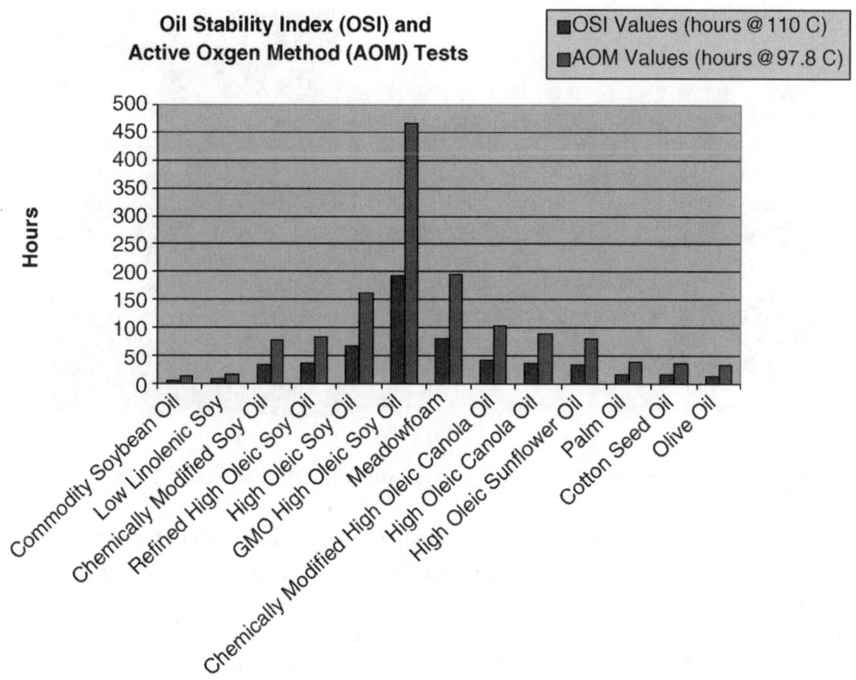

Figure 6.1 OSI values for selected vegetable oils

to be used in biofuels and biobased products. The AAIC has a list of several alternative industrial crops that its members are investigating. Crops most noteworthy in this investigation for lubricants use include cuphea, camelina, canola, castor, lesquerella, peanut, and pennycress. Some of these crops, like camelina, have reached the commercial production stage, and reasonably large acreages are being produced in Montana to accommodate for this growth. Other industrial crops like guayule (why-you-lee) and Hevea (the Brazilian rubber tree, *Hevea brasiliensis*), are used for non-lubricant industrial applications. Guayule is used in producing natural rubber materials that are less allergenic as well as being renewable. It is a desert plant indigenous to the southwest United States and northern Mexico, and grows to an average height of three feet. The species (*Parthenium argentatum*) is considered an industrial crop and the only species other than Hevea that is used for latex production on a commercial scale. It is a natural and economic source for a variety of green products for use in medical devices, consumer products, industrial materials, and renewable energy. Hevea is used as tackifier in biobased and petroleum-based greases.

Guayule is considered a new industrial crop; but, the growing techniques for this plant have been known for some time. Guayule cultivation and harvesting models share much in common with those successfully used on cotton for many years. Currently, guayule-derived natural rubber is marketed as a safe, natural alternative for Type I latex allergy sufferers and in many other medical and industrial applications. Figure 6.2 shows a guayule field (left) and natural rubber extracts (right) [4].

Figure 6.2 Guayule field (left) and natural guayule rubber (right). See Plate 5 for the color figure

Table 6.1 presents the fatty acid makeup of oils from four industrial crops: flax, camelina, babassu, and jojoba (ho-ho-ba), along with the fatty acid makeup of some common conventional crops.

Over the past 30 years, the commercial viability, economic, and technical performance of biobased lubricants have been established. With the increasing demand for petroleum worldwide, and further attention to the carbon footprint of industrial products, there will be a continued development and enhancement of industrial crops for use as chemicals and biobased products. The future technologies will include the old plant oils, the newer genetically enhanced high oleic varieties, and the more sophisticated and economical chemically modified high functioning esters derived from various raw materials. These developments will not completely replace the use of petroleum for industrial and automotive lubricants, but they will

Table 6.1 Fatty acid contents of industrial crops along with canola, soybean and sunflower oils

	Fatty acid content (% of oil)							
Fatty acid	Canola	Soybean	Sunflower	Crambe	Flax	Camelina	Babasu	Jojoba
Caprylic (C 8)							4.5	
Capric (C 10)							7	
Lauric (C 12)							45	
Myristic (C 14)							16	
Palmitic (C 16:0)	6.19	10.44	6.05	2.41	5.12	7.8	7	
Stearic (C 18:0)	0	3.95	3.83	0.4	4.56	2.96	4	
Oleic (C 18:1)	61.33	27.17	17.36	18.36	24.27	16.77	14	10
Linoleic (C 18:2)	21.55	45.49	69.26	10.67	16.25	23.08	2.5	
Linolenic (C 18:3)	6.55	7.16	0.5	5.09	45.12	31.2		
Arachidic (C 20:0)	0	0.5	0.5	0.5	0	0		
Eicosanoic (C 20:1)	1.5	0.5	0	2.56	0	11.99		68.5
Erucic (C 22:1)	0.5	0	0	54	0.88	2.8		
Benenic (C:22)								17
Other FA	2.38	4.79	2.5	6.01	3.8	3.4	2.5	4.5

Figure 6.3 Examples of industrial crops for biobased fuels and lubricants. See Plate 6 for the color figure

capture a significant portion of those markets. Figure 6.3 presents an example of eight industrial crops considered for exploration as a source of non-edible oil for industrial lubricants applications.

Interestingly, many of these industrial crops can produce a much larger volume of oil per acre of land than many of the conventional edible crops. Table 6.2 shows a USDA produced estimate of gallons of oil per acre for each alternative source of biobased oil. When the goal becomes to primarily produce oil crops for non-food industrial uses, these high yielding crops could play an important role as alternatives to petroleum resources [5].

Table 6.2 Estimate of gallons of oil per acre for each alternative source of biobased oil

Crop	Oil content	Gallons of oil per acre
Algae	10–85%	40,000
Oil palm	kernel 50%, fruit 40–70%	760
Pongamia	27–36%	432
Kukui nut	45–65%	380
Euphorbia lathyris	40–48%	315
Jatropha	43–59%	300
Coconut	60–80%	287
Avocado	10–30%	282
Castor bean	40–50%	278
Neem	33–45%	165
Rapeseed	37–50%	127
Pittosporum	1.3 ml fruit^{-1}	125
Peanut	40–55%	113
Sunflower	25–45%	102
Flax	35–40%	51
Soybean	18–20%	48
Copaifera	500 ml tree^{-1} yr^{-1}	8

Table 6.3 Estimated oil yields per unit of surface land area

Crop	kg oil/ha	liters oil/ha	lbs oil/acre	US gal/acre
Avocado	2217	2638	1980	282
Brazil nuts	2010	2392	1795	255
Calendula	256	305	229	33
Camelina	490	583	438	62
Cashew nut	148	176	132	19
Castor seed	1188	1413	1061	151
Coconut	2260	2689	2018	287
Cocoa (cacao)	863	1026	771	110
Coffee	386	459	345	49
Coriander	450	536	402	57
Corn (maize)	145	172	129	18
Cotton	273	325	244	35
Euphorbia	440	524	393	56
Hazelnuts	405	482	362	51
Hemp	305	363	272	39
Jojoba	1528	1818	1365	194
Jatropha	1590	1892	1420	202
Kenaf	230	273	205	29
Linseed (flax)	402	478	359	51
Lupine	195	232	175	25
Macadamia nuts	1887	2246	1685	240
Mustard seed	481	572	430	61
Oats	183	217	163	23
Olives	1019	1212	910	129
Oil palm	5000	5950	4465	635
Opium poppy	978	1163	873	124
Peanuts	890	1059	795	113
Pecan nuts	1505	1791	1344	191
Pumpkin seed	449	534	401	57
Rapeseed	1000	1190	893	127
Rice	696	828	622	88
Safflower	655	779	585	83
Sesame	585	696	522	74
Soybean	375	446	335	48
Sunflowers	800	952	714	102
Tung oil Tree	790	940	705	100

Other sources have documented oil yields per units of surface land area as appear in Table 6.3. These are conservative estimates – crop yields can vary widely [6].

6.2.1 Camelina

Camelina sativa is a potentially important industrial plant for biofuels and biolubricants. Other camelina species include: *Camelina alyssum*, *Camelina microcarpa*, and *Camelina rumelica*. *Camelina sativa* (L.) originated in the Mediterranean and Central Asia. It is an annual or winter annual crop and has stems that become woody at maturity (Figure 6.4).

Figure 6.4 Camelina plant [7]

Camelina has been adapted to the flax-growing regions of the United States' Midwest (Minnesota, North Dakota, South Dakota) and more recently, to the more westerly regions including the state of Montana. Primarily a minor weed in flax, *Camelina* is not a problem in other crops and does not have seed dormancy [8]. However, the adaptation of *Camelina* as a crop has not been widely explored. Similar to the other cruciferous species, it is likely best adapted to cooler climates where excessive heat during flowering is not a problem.

There are several winter annual biotypes available in the germplasm, and it is possible to grow camelina as a winter crop in areas with very mild winters. *Camelina* is short-seasoned (85 to 100 days) so that it can be incorporated into double cropping. The growing of *Camelina* in the state of Montana has now become part of the state's strategy to promote the use of crop oils for fuel and industrial uses. This strategy includes providing cash incentives to farmers who plant *Camelina*. This strategy may be duplicated by other states with a similar climate. Early planting was previously based on simple seed broadcasting. For harvesting, the same equipment used on soybeans has been used.

6.2.1.1 Camelina Seed Composition, Oil Content and Meal Quality

The oil content of the *Camelina* seed ranges from 29 to 39%. *Camelina* appears to be similar in protein content and elemental composition to that of flax (*Linum usitatissimum* L.), with the exception of a higher sulfur content [8]. *Camelina* meal is comparable to soybean meal, containing 45–47% crude protein and 10–11% fiber [9].

6.2.1.2 Use of the Oil

The fatty acids in *Camelina* oil are primarily unsaturated, with only about 12% being saturated. About 54% of the fatty acids are polyunsaturated, primarily linoleic (18 : 2) and linolenic (18 : 3), and 34% are monounsaturated, primarily oleic (18 : 1) and eicosaenoic (20 : 1).

The fatty acid composition of *Camelina sativa* is generally similar to that reported for *Camelina rumelica* [10]. With its low saturated fat content, camelina oil could be considered high quality edible oil. However its high polyunsaturated content, makes this oil susceptible to autooxidation, thus giving it a shorter shelf life. With an iodine value of 144, it is classified as a drying oil. In other uses, camelina oil has replaced petroleum oil in pesticide sprays [11].

Camelina oil is less unsaturated than linseed (flax) oil and more unsaturated than sunflower or canola oils. The balance of saturated vs. unsaturated fats is similar to that of soybean, but camelina contains a significantly higher proportion of C18 : 3 fatty acids. *Camelina* seems to be unique among the species evaluated in having a high eicosenoic acid content in the oil, but the potential value or disadvantage of this is currently unclear.

The erucic acid content is probably too low for use in the same applications as crambe or high erucic acid rapeseed, where high erucic acid content is desired. Most of the camelina lines evaluated contain 2 to 4% erucic acid, which is greater than the maximum (2%) limits for canola-quality edible oil. However, in a preliminary germplasm screen, there are lines with zero erucic acid content, so it is likely that this trait could readily be removed through plant breeding, as it has been with canola.

6.2.2 Babassu

Babassu oil is a clear, light yellow oil extracted from the seeds of the babassu palm tree (*Attalea speciosa*), which is native to Brazil and grows in the Amazon region of South America (Figure 6.5). This tree grows widely and is considered a major industrial and economical resource for the region.

The oil from the babassu tree has many food and non-food uses. Its oil is thus a major reason for growing this tree. Current industrial uses include liquid cleansing products. Babassu oil is mostly saturated and has similar properties to coconut oil. The oil has the same applications and is increasingly being used as a replacement for coconut oil.

Babassu oil is made up of about 70% lipid. The oil comes from the kernels of the babassu palm and due to its saturated and stable fatty acid makeup, it is more of a non-drying oil. It is high in lauric and myristic acids, with melting points close to human body temperature. When applied to the skin, babassu draws heat from the skin to initiate melting. The absorption of heat from the skin creates a cooling sensation. Babassu oil also forms a protective, soothing coat when applied to the skin as well as a pleasant, smooth feeling, without leaving the skin feeling oily. These characteristics create a good combination to make soaps.

The lipids in babassu oil have a fatty acid profile of the following proportions:

Lauric	50%
Myristic	20.0%
Palmitic	11%
Oleic	10%
Stearic	3.5%
Other	5.5%

Figure 6.5 Babassu tree

In 2009, a mixture of babassu oil and coconut oil was used experimentally to partially power one engine of a Boeing 747, in a biofuel trial sponsored by the Virgin Atlantic airline. Like many other vegetable oils, the esters of this oil can be used as a biofuel and show promise to substitute non-renewable oils [12].

6.2.3 Cuphea

Cuphea (family Lythraceae), Figure 6.6, is another prospective industrial crop as a new source of medium-chain fatty acids, especially lauric acid. It contains high levels of other medium-chain fatty acids, like caprylic, capric, lauric, and myristic acids. With the diversity of the cuphea oil fatty acid composition, it could be tailored in a way to produce selected fatty acids under varying growing conditions. At this time, both traditional plant breeding techniques and genetic modification methods are directed toward the development of successful *Cuphea* oilseeds for industrial base materials [13].

Forcella, Gesch and Isbell performed experimental agronomics on *Cuphea* in the Midwestern states of the United States to determine the seed yields, seed oil contents, and fatty acid profiles plus other information. Accordingly,

> *... in the absence of drought, cuphea grew well vegetatively at most sites, but in the US, seed yields tended to be higher in Minnesota (which borders Canada and is farther north) than in Iowa [13].*

The oil content of seeds varies inversely with growing temperatures, with the seed oil content ranging from 28% at higher temperatures to 33% at lower temperatures. The capric acid content ranged from 67 to 73% of total oil, and was always highest in the colder, northern-most sites. In

Figure 6.6 Cuphea. See Plate 7 for the color figure

the United States, *Cuphea* can become a source of lauric acid. It is currently imported from palm-producing regions [13].

6.2.4 Castor

Castor oil is a vegetable oil obtained from the castor seed of the castor plant (*Ricinus communis*) – Figure 6.7. Castor oil is a colorless to very pale yellow liquid with mild or no odor or taste. Its boiling point is 313 °C (595 °F) and its density is 0.97 g/mL.

The sulfonated or sulfated castor oil, also referred to as Turkey Red Oil, completely disperses in water. As a result, it has uses in body wash products, soaps and detergents. Adding sulfuric acid to pure castor oil produces sulfated oil. It was the first synthetic detergent as a replacement for ordinary soap, while also having other uses in lubricants and softeners.

Harvesting castor beans involves health risks because the plant surface has allergenic compounds with the potential to cause nerve damage. The castor seed contains ricin, a highly toxic protein. Castor is grown in India, Brazil, and China. It is a major source of hydroxyl fatty acid. The health hazards associated with its harvest, however, has resulted in a conscious search for alternative means of producing hydroxyl fatty acids. Some researchers are also attempting to genetically modify the castor plant to prevent the synthesis of ricin.

In the food industry, castor oil (food grade) is used in food additive, flavorings, and candy or chocolate. Castor oil has many medicinally beneficial properties for human use that have

Figure 6.7 Castor plant. Courtesy of the Schundler company. See Plate 8 for the color figure

been used in home remedies for generations. The United States Food and Drug Administration (FDA) classifies castor oil to be "Generally Recognized As Safe and Effective" (GRASE) for use as a laxative. Some derivatives of castor oil such as undecylenic acid are also FDA-approved for use on skin disorders or skin ailments. For other purposes, it has properties that inhibit the growth of mold and thus can be used as mold inhibitor, for instance, in packaging.

Ricinoleic acid, which is a main component in castor oil, has anti-inflammatory properties. Like jojoba oil, castor oil too penetrates into the skin because of its chemical structure. Castor isostearate succinate is a polymeric mixture of esters with isostearic acid and succinic acid used for skin conditioning in products such as shampoo, lipstick, and lip balm [14].

Therapeutically, modern drugs are rarely given in a pure chemical state, so most active ingredients are combined with additives. Castor oil, or a castor oil derivative such as Cremophor EL (polyethoxylated castor oil, a non-ionic surfactant), is added to many modern drugs, from antifungal products to topical skin treatments.

In addition to having medicinal value as found in over-the-counter drugs, castor oil was used extensively in early rotary engines common in early airplanes. It was also used as engine oil in methanol-fueled glow plug type model airplane engines.

Castor oil continues to be a source of ricinoleic acid and thus an important component of food, medicine, and lubricants. Due to its high viscosity it can be used as a viscosity modifier, and due to its low pour point it can be used as a cold temperature flowability improver. Until transgenic crops containing ricinoleic acid are developed, castor oil will play a role in the production of numerous products. The fatty acid profile of castor oil is presented in Table 6.4 [15].

Table 6.4 The ranges for fatty acid composition of castor oil

Fatty acid composition of castor oil			
Fatty acid	Range of percentages		
Ricinoleic acid	85	To	95%
Oleic acid	2	To	6%
Linoleic acid	1	To	5%
Linolenic acid	0.5	To	1%
Stearic acid	0.5	To	1%
Palmitic acid	0.5	To	1%
Dihydroxystearic acid	0.3	To	0.5%
Others	0.2	To	0.5%

The oxidative stability of castor oil along with those of other vegetable oils is presented in Table 6.9. Of the oils presented, castor oil shows the highest oxidation stability index of 105 hours.

6.2.5 *Rice Bran*

Rice (*Oryza sativa* Linn.) bran is a byproduct of rice milling. During the process of milling rice to produce polished rice, rice bran is obtained from the outer layers of the brown (husked) rice kernel. Whole rice grain comprises (on a dry basis) 70–72% endosperm, 20% hull, 7–8.5% bran, and 2–3% embryo.

Rice bran comprises pericarp, tegmen (layer covering endosperm), aleurone, and sub-aleurone, Figure 6.8. The rice bran contains 15–23% oil, which is one of the most nutritious oils

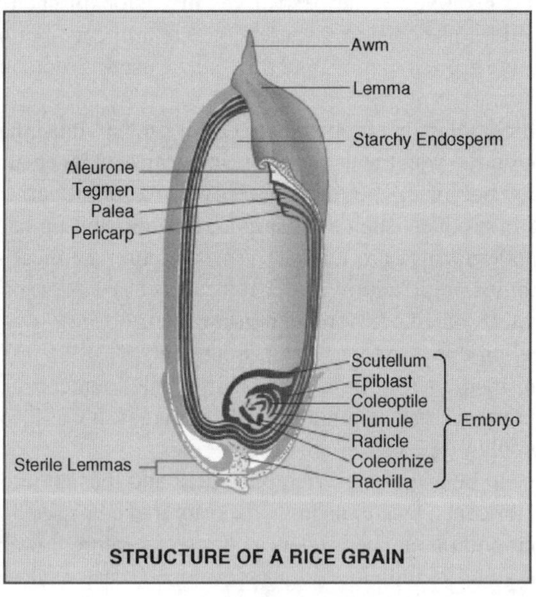

Figure 6.8 Rice bran. Photo courtesy of www.thailandonline.com

because of its favorable fatty acid composition. Lipase enzyme in rice bran adversely affects the storage quality and the subsequent industrial use of bran. The enzyme becomes active right after milling and the rate of free fatty acids formation is as high as 5–7% per day depending on environmental conditions. It can go up to 70% after storage of 1 month [16].

6.2.6 Jatropha

Jatropha curcus, Figure 6.9, is a drought-resistant perennial, growing well in marginal/poor soil. This propensity to survive in marginal soil provides the opportunity to grow fuel and lubricants in countries with a poor petroleum supply and poor or dry soil. It is easy to establish, grows relatively quickly, and lives producing seeds for 40 years.

The *Jatropha* plant produces seeds with an oil content of 37%, almost twice of that of soybeans and nearly the same amount of camelina. The oil is a fuel that burns with a clear, smoke-free flame. Upon processing this oil into biodiesel (through esterification), it is increasingly being used as a fuel by transport and energy companies. The by-products are pressed meal, which is used as organic fertilizer. The oil also contains an effective and natural insecticide. Jatropha is a highly adaptable species, but its strength as a crop comes from its ability to grow on very poor and dry sites.

Jatropha, from the Greek words ***jatros*** meaning physician and ***trophe*** meaning nutrition, is also called **'physic nut'.** It is a genus of approximately 175 succulent plants, shrubs and trees from the family Euphorbiaceae. *Jatropha* is native to Central America and has become

Figure 6.9 Jatropha plant. Photo courtesy of the Centre for Jatropha Promotion and Biodiesel India. See Plate 9 for the color figure

naturalized in many tropical and subtropical areas, including India, Africa, and North America [17].

In some countries it is used medically for health conditions including cancer, hemorrhoids, snakebite, paralysis, and dropsy. Depending on soil quality and rainfall, oil can be extracted from the jatropha nuts after two to five years. The annual nut yield ranges from 0.5 to 12 tons. The kernels consist of oil to about 60%; this can be used for lubricants or transformed into biodiesel fuel through esterification.

Aviation fuels may be more widely substituted with biofuels such as jatropha oil than fuels for other forms of transportation. On 30 December 2008, Air New Zealand flew the first successful test flight with a Boeing 747 running one of its four Rolls-Royce engines on a 50 : 50 blend of jatropha oil and jet A-1 fuel [18]. Subsequently, Air New Zealand and Houston-based Continental Airlines have run tests, further demonstrating the viability of jatropha oil as a jet fuel. Japan Air also planned test flights in 2009.

The following provides the proximate fatty acid profile of jatropha oil:

Free fatty acid	<2.0% w/w
Water content	<1000 ppm
Phosphorus	<20 ppm w/w
Sulfur	<50 ppm
Iodine value (mg I_2/100g	<120
Saponification number	>190 (mg KOH/g)
Specific gravity	0.840–0.920

The fatty acid profile of jatropha oil includes:

Myristic acid	0.38%
Palmitic acid	16.0% max.
Palmitoleic acid	1–3.5%
Stearic acid	6–7.0%
Oleic acid	42–43.5%
Linoleic acid	33–34.4%
Linolenic acid	>0.80%
Arachidic acid	0.20%
Gadoleic acid	0.12%

6.2.7 Neem

Another potential source of industrial oil comes from the neem tree which grows in dry climates. The neem tree *Azadirachta indica* A. Juss. (Meliaceae) is an evergreen tree of the mahogany family that is native to India and Burma (Figure 6.10). It is found in tropical and subtropical climates, and can withstand extremely dry conditions, but also tolerates sub-humid conditions. Neem trees are fast-growing and can grow up to 35 m tall, and although evergreen, they will lose their leaves in times of severe drought. They have wide, spreading branches and relatively short trunks. One tree can produce millions of flowers, and in one flowering cycle, a

Figure 6.10 Branches of neem tree and seeds. Indian Biofuels Awareness Centre, retrieved April 10, 2010 from www.svlele.com/biofuel.htm [27]

mature tree may produce many thousands of seeds. Seeds are small and round to oval in shape, with oil content ranging from 20 to 33%, depending on the variety.

This tree has adapted to a wide range of climates and thrives well in hot weather, where the maximum shade temperature is as high as 49 °C (120 °F) and tolerates cold up to 0 °C (32 °F) on altitudes up to 1500 m.

The neem tree can grow on almost all types of soils including clay, saline and alkaline soils, with pH up to 8.5, but does well on black cotton soil and deep, well-drained soil with good subsoil water. It needs little water and plenty of sunlight. It grows naturally in areas where the rainfall is in the range of 450 to 1200 mm. A Neem tree normally begins to bear fruit between 3 and 5 years and becomes fully productive in 10 years. A mature tree produces 30–50 kg of fruit each year. It has a productive life span of 150–200 years [20].

The average percent of fatty acid content of neem is reported as oleic acid 45.6%, linoleic acid 16.8%, palmitic acid 17.21%, stearic acid 15.2%, and linolenic acid 1.3%. Table 6.5 shows the fatty acid profile of Neem seeds.

Table 6.5 Average fatty acids profile of neem seeds

Kernel content%	Fatty acid%				
	Oleic	Linoleic	Palmitic	Stearic	Linolenic
30	45.73	18.44	18.21	15.70	1.33
25	45.73	18.72	17.93	15.03	1.25
20	44.99	16.55	18.33	17.02	1.49
Average	45.55	16.77	17.21	15.23	1.33

Figure 6.11 Karanja (pongam) tree

6.2.8 Karanja (Pongam)

Karanja is a city in the state of Maharashtra, India. The Karanja tree (*Pongamia glabra*) is widely found in most of tropical Asia, Figure 6.11. It is grown ornamentally in gardens and the roadside for its fragrant flowers, and as a host plant for lac insects. Karanja can grow on most soil types ranging from stony to sandy to clay, including verticals. It does not do well on dry sands however. It is common along waterways or seashores, with its roots in fresh or salt water as it is highly tolerant of salinity.

Lac is the scarlet resinous secretion of a number of species of plant-sucking insects, a few of which produce natural products (e.g. cochineal and crimson). Large numbers of these tiny insects colonize branches of suitable host trees and secrete the resinous pigment. The coated branches of the host trees are cut and harvested as sticklac. Once extracted the lac is used on violins and in other varnishes and is soluble in alcohol. This type of lac has had uses in the finishing of various wood products like the violin and rifle stock. Seedlac which still contains 3–5% impurities is processed into shellac by heat treatment or solvent extraction. Shellac is then used as lacquer; although nowadays these products are synthesized from other materials. Shellac is also used as a glaze on many edible products including candy [20].

The tree is hardy and drought resistant and in India, for example, it grows to a height of about 1 m. The seeds contain the karanja oil, a red brown, thick, non-drying, non-edible and bitter oil, 27–36% by weight. The oil has a high content of triglycerides, and its disagreeable taste and odor are due to bitter flavonoid constituents, pongamin and karanjin. Both the oil and residues are toxic. Still, the meal is used as poultry feed. Dried karanja tree leaves are used in stored grains to repel insects. Twigs are used in some countries as a chew-stick for cleaning the teeth.

The fatty acid composition of karnenja oil is as follows:

Palmitic	3.7–7.9%
Stearic	2.4–8.95%
Arachidic	2.2–4.75%
Behenic	4.2–5.35%
Lignoceric	1.1–3.55%
Linoleic	10.8–18.3%
Eicosenoic	9.5–12.4%
Oleic	44.5–71.3%

Destructive distillation of the wood yields, on a dry weight basis: charcoal 31.0%, pyroligneous acid 36.69, acid 4.3%, ester 3.4%, acetone 1.9%, methanol 1.1%, tar 9.0%, pitch and losses 4.4%, and gas 0.12 m^3/kg. Manurial values of leaves and twigs are respectively: nitrogen 1.16, 0.71; phosphorus (P_2O_5), 0.14, 0.11; potash (K_2O), 0.49, 0.62; and lime (CaO), 1.54, 1.58 [21].

6.2.9 Poppy

The **opium poppy**, *Papaver somniferum*, Figure 6.12, is the type of poppy from which opium and many refined opiates, including morphine, thebaine, codeine, papaverine, and noscapine, are extracted. Poppy is grown for many uses including for pain management in pharmaceuticals and in illicit drugs. The seeds of the poppy are widely used in and on many food items such as bread, bagels, muffins, and cakes. The seeds can be pressed to form poppy seed oil. The oil from poppy seed is clear with low oxidation stability due to the presence of high level of linoleic acid. The oil has many uses, including as an artist's oil for oil painting [22]. The approximate fatty acid content of poppy seed oil is as follows:

Palmitic acid	12%
Stearic acid	3%
Oleic acid	20%
Linoleic acid	65%

Figure 6.12 Poppy seed [23] and poppy field. See Plate 10 for the color figure

6.2.10 Sesame

Sesame is another crop with significant use in foodstuffs. As with most plant-based condiments, sesame oil contains magnesium, copper, calcium, iron, zinc and vitamin B6. These are useful components with health benefits. For example, copper provides relief for rheumatoid arthritis and magnesium supports vascular and respiratory health. Similarly, calcium helps prevent colon cancer, osteoporosis, migraines, and premenstrual syndrome. Zinc promotes bone health.

The use of such oils for industrial applications may be prohibitive due to the high food value. But, understanding their physiochemical properties is helpful in their use as property enhancers or transesterification with other vegetable oils.

Sesame oil is composed of the following fatty acids [24]:

Table 6.6 Approximate fatty acid profile of sesame seed oil

Fatty acid	Carbon chain	Minimum	Maximum
Palmitic	C16 : 0	7.0%	12.0%
Palmitoleic	C16 : 1	Trace	0.5%
Stearic	C18 : 0	3.5%	6.0%
Oleic	C18 : 1	35.0%	50.0%
Linoleic	C18 : 2	35.0%	50.0%
Linolenic	C18 : 3	Trace	1.0%
Eicosenoic	C20 : 1	Trace	1.0%

Source: http://www.essentialoils.co.za/sesame-oil-analysis.htm

6.2.11 Jojoba

Jojoba oil is the waxy liquid produced in the seed of the jojoba (*Simmondsia chinensis*) plant (Figure 6.13). It is a shrub native to southern Arizona, southern California and northwestern Mexico. The oil makes up approximately 50% of the jojoba seed by weight [25].

Jojoba oil is a mixture of long chain wax esters, ranging from 36 to 46 carbon atoms in length. Each molecule consists of a fatty acid and a fatty alcohol joined by an ester bond. 98% of the fatty acid molecules are unsaturated at the 9th carbon–carbon bond (omega-9) [26].

The approximate percentages of fatty acids in jojoba oil are as follows:

Eicosenoic	66–71%
Docosenoic	14–20%
Oleic	10–13%

The crude unrefined jojoba oil is a clear, golden liquid at room temperature, with some fatty odor. The melting point of jojoba oil is approximately 10 °C and the iodine value is approximately 80.

Jojoba oil is relatively shelf-stable when compared with other vegetable oils. It has an oil stability index of approximately 60. This means that it is more shelf-stable than oils of safflower oil, canola oil, and soybean oil but less than castor oil, macadamia oil, and coconut oil.

Figure 6.13 Jojoba plant

6.2.11.1 Uses of jojoba

One the most common use of jojoba oil is in cosmetics and personal care products. Due to its natural ability to penetrate the skin it has especial suitability for skin and hair care products. Jojoba oil on the skin absorbs moisture from the air making it an effective moisturizer. Derivatives of jojoba, including jojoba esters, isopropyl jojobate and jojoba alcohol, are particularly widely used in this context. Approximately ninety percent of fatty acid chains are ricinoleic acid. Oleic and linoleic acids are the other significant components [27].

Jojoba oil is also used as a replacement for whale oil and its derivatives, such as cetyl alcohol. The ban on importing whale oil to the United States in 1971 led to the discovery that it is "in many regards superior to sperm oil for applications in the cosmetics and other industries" [28].

Jojoba oil has properties that enable it to work as a fungicide, and so it can be used for controlling mildew. The oil is edible, but acaloric and non-digestible, meaning the oil will pass through the intestines unchanged and can cause an unpleasant result called steatorrhea.

Methyl esters of jojoba have been explored as a cheap, sustainable biodiesel.

6.2.12 Coconut

Coconut oil has been a source of food for many years. The oil is very stable because it has a high level of saturated fats. The oil is solid at temperatures under degrees 24 °C (76 °F). **Coconut oil** is extracted from the kernel or meat of the coconut, harvested from the coconut palm (*Cocos nucifera*).

Coconut oil is absorbed in the skin making the skin feel soft without feeling greasy. It has skin-healing effects and is used to soothe and heal wounds and rashes, and will reduce chronic

Figure 6.14 (a) Coconut tree, (b) coconuts and coconut oil, and (c) illustration of coconuts [23]. Reproduced from Mt Banahaw Health Products website www.coconutoil.com. See Plate 11 for the color figure

skin inflammation within days. In the making of soaps, the soap has a more pleasant aroma, unlike the smell of tallow or some other vegetable oils. It has a nice fresh smell and is one of the most popular oils used in soap making.

Coconut oil is a natural white color useful for making pure white-colored soaps. Coconut oil has long been regarded as one of the best base oils for making soap, shampoo, and detergents. The antibacterial, antiviral, and antifungal properties of the medium chain fatty acids in coconut oil make the soap an effective disinfectant. This natural germ-fighting action eliminates the need to add antiseptic chemicals to the soap.

Like many other oils, the use of coconut oil for making biofuel and estolides for lubricants. Methyl esters of coconut can be produced like other vegetable oils presenting a more stable ester, but have higher pour point. Due to the ready availability of coconut in many parts of the world, there are research activities involving industrial and automotive lubricants such as engine oils. Figure 6.14 shows coconut and coconut oil as well as a coconut tree. Table 6.7 shows the approximate fatty acid profile for coconut oil [29].

Table 6.7 Approximate fatty acid content of coconut oil

Fatty acid	Chain length	Approximate value
Caprolic acid	C6	0.20%
Caprylic acid	C8	4.80%
Capric acid	C10	4.80%
Lauric acid	C12	54.50%
Myristic acid	C14	18.80%
Palmitic acid	C16	8.30%
Stearic acid	C18 : 0	2.80%
Oleic acid	C18 : 1	5.00%
Linoleic acid	C18 : 2N6	0.80%

Figure 6.15 Lesquerella. Courtesy Smithsonian Institution, Department of Systematic Biology-Botany

6.2.13 Lesquerella

Lesquerella fendleri is a wild, yellow-flowered member of the mustard family native to Arizona, New Mexico, Colorado, Utah, Texas, and Mexico (Figure 6.15). Three hydroxy fatty acids in lesquerella oil, including lesquerolic, densipolic, and auricolic, are similar to ricinoleic acid, the main fatty acid in castor oil.

The lesquerella oil-bearing seed contains high amounts of the hydroxy fatty acid (HFA), lesquerolic acid (C20:1-OH), suitable as a raw material for many types of industrial applications. Lesquerella oil can be considered a possible replacement for castor oil as a source of hydroxy fatty acid for the production of lubricants and plastics. Lesquerella oil could complement or replace castor oil imports due to the hazards associated with the harvest and processing of castor oil [30].

With minor modification, existing farm equipment can be used for planting and harvesting the lesquerrella crop. Genetic and germplasm improvement studies are also being conducted and cover germplasm evaluation, selection, and hybridization.

Lesquerella oil has a red-to-brown color, which becomes a problem when attempting to use it as a substitute for castor oil, particularly in cosmetic products. Efforts are underway to produce genetically modified versions of the lesquerella oil that yields lighter color oil. Both the oil and the gums from *Lesquerella* have high value as base stock for industrial uses. The presence of hydroxy fatty acids [31] provide for great cold temperature flowability, which is desirable for industrial lubricants uses.

6.2.14 Hemp

Hemp is a fibrous weed of the *Cannabis sativa* species with dark green leaves and grown worldwide. The use of hemp for textile, rope making, its oil for food, and its leaves for

Figure 6.16 Hempseed

medicinal and other uses have been known for millennia. The plant is a tall weed and is grown worldwide. Aside from its use in hemp seed oil, the plant itself grows rapidly (four times faster than trees). It's promoted as a renewable biomass than can be grown without fertilizer or pesticides. Figure 6.16 shows hemp seeds.

Cannabis sativa can be separated into two categories:

- Hemp (drug type): the leaves are rich in THC (tetrahydrocannabinol), do not contain its precursor CBD (cannabidiol), and are used for its psychotropic properties.
- Hemp (fiber type): contains very low levels of THC and does contain CBD.

Hemp seed oil has a dry texture and on the skin it does not feel oily. This feature is useful in manufacturing beauty products. The approximate fatty acid content of hemp seed oil appears in Table 6.8.

6.2.15 Flaxseed oil

Flaxseed oil is derived from the seeds of the flax plant (*Linum usitatissimum* L.). Flaxseed oil and flaxseed contain alpha-linolenic acid (ALA), an essential fatty acid that appears to be

Table 6.8 Approximate fatty acid profile of hemp seed oil

Fatty acids of hemp seed oil	Percent of total
Alpha linolenic	18%
Gamma linolenic	2.70%
Oleic	11%
Linoleic	57%
Palmitic	6.50%
Stearic	2.50%
Other	2.30%

beneficial for curing heart disease, inflammatory bowel syndrome, arthritis, and other health conditions. Flaxseed (Figure 6.17) also contains a group of chemicals called lignans that may play a role in the prevention of cancer and show cholesterol-lowering properties.

The major fatty acids of flaxseed are:

Palmitic	5%
Oleic	20%
Linoleic	53.25%
Other	21.75%

6.2.16 *Safflower*

Safflower (*Carthamus tinctorius* L.) is a highly branched, herbaceous, thistle-like annual, usually with many long sharp spines on the leaves. Plants can reach heights of 30 to 150 cm with globular brilliant yellow, orange or red flowers that bloom in July. Each branch will usually have from one to five flower heads containing 15 to 20 seeds per head. Safflower has a strong taproot which enables it to thrive in dry climates, but the plant is prone to frost damage.

The safflower plant (Figure 6.18) has been cultivated mainly for the vegetable oil extracted from its seeds. Efforts have been underway to genetically modify safflower to create insulin.

Figure 6.17 Flaxseed. Photo courtesy of Claridges Organic New Zealand

Figure 6.18 Safflower plant. With kind permission from Springer Science+Business Media: *Journal of the American Oil Chemists Society*, Jojoba oil analysis by high pressure liquid chromatography and gas chromatography/mass spectrometry, 54, 1977, Spencer, GL, *et al.*

Safflower may have varieties with high monounsaturated fatty acid (more oxidatively stable) or varieties high in polyunsaturated fatty acid (less oxidatively stable). The latter is used as a substitute for linseed in artist's oil for oil painting. Dried safflowers are used as a natural textile dye for coloring textiles.

The chemical properties and fatty acid profile of safflower oil is presented in approximation below:

Peroxide value	5 max. (mg/kg)
Iodine value	138–150
Refractive index	1.47–1.48 @ 200 °C
Acid value	0.3 max
Smoke point	200 °C min.
Saponification value	185–198
Approximate fatty acid profile	
Palmitic C16 : 0	4–7%
Stearic C18 : 0	0.5–2%
Oleic C18 : 1	8–12%
Linoleic C18 : 2	80–84%

6.3 Future and Industrial Crops

With the growing worldwide demand for petroleum, the global petroleum supply has become more constrained. Renewable oil stocks are again gaining favor as a potential solution throughout our increasingly industrialized world. Conventional oil crops domesticated over many generations and produced in well established and heavily mechanized operations will no doubt play an important role in this transition. Widely known conventional crops will be the mainstay of this move to renewable oil stocks, especially in the earlier stages. Continued improvement in agronomics and advances in the farming machinery and farming technology will result in continued increased yields on the same acreages of land.

Table 6.9 Physiochemical, rheological, and tribological properties of selected vegetable oils

Oil	OSI (hours)	TAN	Flash point (PM)	Flash point (COC)	Fire point (COC)	Pour point (°C)	Cloud point (°C)	Viscosity @ 40	Viscosity @ 100	Viscosity index	4 Ball Wear	Pin and Vee (ref. load lbs)
Apricot kernel	23.42	0.2844	284.5	324	348	−16	−10.8	36.49	8.202	210	0.615	1732
Avocado	18.53	0.185		320	348	−3	−0.2	39.26	8.432	199	0.609	1975
Babassu	57.8	N/A				N/A	N/A	28.65	6.133	170	0.586	1706
Castor	105.13	0.252		300	320	−28	N/A	249.5	19.02	85	0.633	1674
Coconut	75.38	N/A				N/A	N/A	27.8	5.947	167	0.504	1738
Corn	3.73	0.198		324	346	−15	−10.2	32.58	7.72	220	0.628	1997
Cottonseed	4.35	0.13	262			−6	−3.7	34.23	7.911	215	0.588	1812
Flaxseed	1.17	0.8399	268	322	348	−12	−7.4	27.35	7.112	243	0.639	1622
Grapeseed	2.83	0.229	248			−12	−6.9	33.28	7.858	220	0.623	1736
Hempseed	0.10	1.6488	248			−15.8	−28	26.71	6.972	242	0.608	1556
Jojoba – refined	42.15	0.13		304	330	9	9	25.1	6.519	234	0.630	1673
Jojoba – golden	38.3	0.752		304	330	10.7	8	24.82	6.452	233	0.606	1558
Lard		N/A				N/A	N/A	N/A	8.543	N/A		1676
Macadamia	6.87	0.126	276			−5	−1.9	39.24	8.441	200	0.594	1797
Oleic acid						3	5.9	19.05	4.778	186		1341
Olive	5.08	0.132		316	342	−6	−5.4	37.56	8.242	203	0.616	1683
Palm kernel		N/A				N/A	N/A	31.96	6.606	169		1622
Palm	21.52	N/A				N/A	N/A	41.77	8.56	189		1726
Poppyseed	17.86	0.151	256			−18	−15.5	30.52	7.46	226	0.601	1908
Ricebran	20.82	0.194	248			−9	−3.9	36.49	8.177	208	0.581	1549
Ricinoleic acid	117.1		253			−19	−5.5				0.519	1277
Safflower	17.98	0.1268		322	350	−22	0.4	37.9	8.325	206	0.634	1660
Sesame	5.8	0.136	266			−9	−5.7	34.1	7.923	216	0.49	1842
Soy	17.67	0.1602	292	328	346	−9	−5.1	31.08	7.552	226	0.601	1835
Soy HOBO (08-204)		0.2346	248			−12	−9.9	39.12	8.492	203	0.608	1768
Sunflower	10.23	0.132	272			−15	−9.9	38.58	8.453	205	0.621	1864
Walnut	16.48	0.1269		322	346	−19	−14.5	29.91	7.441	232	0.584	1887

As seen through the past decade, however, the increased demand for renewable oil stocks for biofuels production has also piqued the interest of both the research and finance communities in developing suitable alternative oil crops. The growing human population creates additional demands on our food supply, requiring more attention to non-edible oil crops. Numerous indigenous, non-edible industrial oil crops can be grown efficiently in a variety of regions, with local production technology. By producing these crops on marginal lands, the competition with the food supply will be avoided.

Eventually, genetic modifications and transgenic technologies will provide non-edible oil crops with improved yields and enhanced performance for specific use as fuels and lubricants. In the very near future, conventional crops will provide the raw materials for many industrial applications. At the same time, the production of non-edible industrial crops will expand to provide oil stocks for biofuels and biolubricants. Ultimately, in a world where the limited petroleum supply will push prices above the reach of small, developing countries, the indigenous oil crops will fill this need. Industrial crops hold the potential for decentralized lube and fuel production capabilities in an environment where the "grow local, buy local" philosophy encompasses food, lubes, and fuels. In the future, industrial oil crops with genetic improvements targeting varieties of specific oil properties and growing regions will serve as a viable petroleum replacement. The outcome will be oils that are well suited to crop-based lubricants, chemicals, and feedstock for biopower, and biofuels.

Future research will need to focus on not only the chemical composition of crop oils, but also on their physical, rheological, and tribological properties. With the availability of such information, the lubricant and grease formulators will be able to choose base oils that provide the most economical and effective final product. Table 6.9 shows a number of potential industrial crops with their selected chemical properties like OSI values, physical properties like flash and fire points, rheological properties like viscosity and viscosity index; and tribological properties like wear and extreme pressure test results.

References

1. Salunkhe, J.K., Chavan, J.K., Adsule, R.N., and Kadem, S.S. (1992) *World Oilseeds: Chemistry, Technology, and Utilization*, Van Nostrand Reinhold, New York.
2. Kinney, J.A. (2005) US Patent 6949698. Gene combinations that alter the quality and functionality of soybean oil.
3. Honary, L.A.T. (1995) Performance of Selected Vegetable Oils in ASTM Hydraulic Tests. SAE Technical Papers, Paper 952075.
4. Yulex Corporation. Retrieved 18 April 2010 from http://www.yulex.com/cropscience/growing.php.
5. US Department Energy Biomass R&D Newsletter, April 2009.
6. Indian Biofuels Awareness Centre, retrieved 12 April 2010 from: www.svlele.com/biofuel.htm.
7. http://www.physorg.com – retrieved 12 April, 2010.
8. Robinson, R.G. (1987) Camelina: A useful research crop and a potential oilseed crop. *Minnesota Agr. Expt. Sta. Bul.* 579–1987. (AD-SB-3275).
9. Korsrud, G.O., Keith, M.O., and Bell, J.M. (1978) A comparison of the nutritional value of crambe and camelina seed meals with egg and casein. *Can. J. Anim. Sci.* **58**, 493–499.
10. Umarov, A.U., Chernenko, T.V., and Markman, A.L. (1972) The oils of some plants of the family cruciferae. Khimiya Prirodnykh Soedinenii, USSR – pp. 24–27.
11. Robinson, R.G. and Nelson, W.W. (1975) Vegetable oil replacements for petroleum oil adjuvants in herbicide sprays. *Econ. Bot.* **29**, 146–151.
12. http://news.bbc.co.uk/2/hi/7261214.stm – retrieved 2 February 2011-02-05.

13. Forcella, F., Gesch, R.W., and Isbell, T.A. (2003) Crop Ecology, Management & Quality. Seed Yield, Oil, and Fatty Acids of Cuphea in the Northwestern Corn Belt.
14. http://www.bkherb.com/HerbalExtracts/Castor_Oil/ retrieved April 4, 2010.
15. http://en.wikipedia.org/wiki/Castor_oil – Retrieved April 4, 2010.
16. Indian Biofuels Awareness Centre, retrieved 10 April 2010 from www.svlele.com/biofuel.htm.
17. www.circlebio.com/jatropha_biodiesel.htm – Retrieved April 4, 2010.
18. Kanter, J. (2008). Air New Zealand flies on engine with jatropha biofuel blend. *New York Times*.
19. Muñoz-Valenzuela, S., Ibarra-López, A.A., and Marian, L. (2007) Neem tree morphology and oil content, in *Issues in New Crops and New Uses* (eds J. Janick and A. Whipkey), ASHS Press, Alexandria, VA.
20. Indian Biofuels Awareness Centre, retrieved April 10, 2010 from www.svlele.com/biofuel.htm.
21. Duke, J.A. (1983) Handbook of Energy Crops. www.tstanes.com/oil_cakes.php – Retrieved April 4, 2010.
22. Sengupta, A. and Mazumder, U.K., Department of Pharmacy, Jadavpur University, Calcutta-32, India. http://www.hort.purdue.edu/newcrop/duke_energy/papaver_somniferum.html, – Retrieved April 4, 2010.
23. Magnusson, G., Hermansson, G., and Leissner, R. (eds) (1998) *Vegetable Oils and Fats*, Karlshamns AB, Sweden.
24. Esoteric Oils CC. Fatty acids found in sesame oil. Essential oils, retrieved 10 April 2010 from: http://www.essentialoils.co.za/sesame-oil-analysis.htm.
25. Undersander, D.J., Oelke, E.A., Kaminski, A.R., *et al.* (1990). Jojoba. Alternative Field Crops Manual. www.demec.ufmg.br/disciplinas/eng032-BL/artigo.pdf – Retrieved April 4, 2010.
26. Spencer, G.F., Plattner, R.D., and Miwa, T. (1997) Jojoba oil analysis by high pressure liquid chromatography and gas chromatography/mass spectrometry. *Journal of the American Oil Chemists' Society*, **54**(5).
27. http://www.shea-baby.com/ingredients.html – Retrieved April 4, 2010.
28. http://en.wikipedia.org/wiki/Jojoba_oil – Retrieved April 4, 2010.
29. Wilderness Family Naturals, retrieved 10 April 2010 from http://www.wildernessfamilynaturals.com/virgin_coconut_oil_traditional.htm.
30. Sierig, D.A., Thompson, A.E., and Nakayama, F.S. (1992) Lesquerella commercialization efforts in the United States. *Industrial Crops and Products*, **1**(2–4), 289–293.
31. Jenderek, M.M., Dierig, D.A., and Isbell, T. (2009) Fatty acid profile of Lesquerella germplasm in the National Plant Germplasm System collection. *Industrial Crops and Products*, **29**(1), 154–164.

7

Biobased Lubricants Technology

The technology for biobased lubricants, initially referred to as biodegradable lubricants, was based on vegetable oils with minor chemical modification and performance enhancing additives. Since vegetable oils generally face inherent challenges when it comes to industrial lubricant uses, their performance properties must therefore be carefully studied. Soybean oil, for example, shows a significant lack of oxidation stability, with an oil stability index (OSI) value of about 7 hours. In one case, this oil was partially hydrogenated to improve its oxidation stability and then winterized to improve its pour point performance. Honary, in a 1998 patent # 5,972,855, describes the formulation of a tractor hydraulic oil made from partially hydrogenated and winterized soybean oil and additive package containing antioxidants, and pour point depressants [1].

Perhaps the most important development for US biobased lubricants was the introduction of high oleic acid soybeans by the DuPont Corporation in the early 1990s. This genetically enhanced soybean had oil with a fatty acid profile considerably superior to conventional soybean oils. Originally designed for frying applications, this oil showed an OSI value of 192 or about 27 times more stable than conventional soybean oils (about 7 hours OSI). This helped in the creation of a number of highly successful lubricants and greases. Table 7.1 shows the physical as well as rheological performance differences of conventional soybean oils with the high oleic soybean oil. In the 1990s, the Lubrizol Corporation had built its additive and lubricants technology based on high oleic and ultra high oleic sunflower oils. Still today, for many industrial lubricants applications, high oleic soybean oil, high oleic sunflower oil, and high oleic canola oils are base oils of choice.

The following section provides a review of the practical consequences of the deficiency of oxidation stability in vegetable oils.

7.1 Determination of Oxidation Stability

From a practical point of reference, oxidation stability refers to the ability of the oil to maintain its properties, mainly its viscosity, when exposed to specific operating conditions. Since vegetable oils have been a main ingredient in the food industry, the majority of methods dealing

Biobased Lubricants and Greases: Technology and Products, First Edition. Lou A.T. Honary and Erwin Richter

Table 7.1 Viscosity, viscosity index, and pour points of selected oils and identical hydraulic fluids utilizing soybean oil and mineral oil base fluids

Description	ASTM D6749 Pour point (°C)	ASTM D445 Viscosity @ 40 °C	ASTM D445 Viscosity @ 100 °C	ASTM D2270 Viscosity index
High oleic soy oil	−16	31.19	8.424	200
Crude conventional soy	−6	31.69	7.589	222
Mineral oil ISO VG 100	−50	20.58	3.684	28
Mineral oil ISO VG	−32	96.21	9.040	53
Mineral oil blend of 57%:43% (of ISO VG 100 and 500)		37.95	5.295	53
Hydraulic fluid with crude conventional soybean oil		32.26	7.592	217
Hydraulic fluid with high oleic soybean oil		39.14	8.412	199
Hydraulic fluid with mineral oil blend		25.24	4.248	46

with their oxidation stability have been created through the efforts of the food industry, and by association, the chemical industry. Oxidation stability methods used in the lubricants and grease industry are based on petroleum and its derivative oils and are *not often suitable for determining* the stability of vegetable oils. Biobased lubricants researchers and developers rely on using standards created by the American Oil Chemists Society (AOCS), and further, the modified versions of standards created by the American Society for Testing and Materials (ASTM). Others at the University of Northern Iowa National Ag-Based Lubricants (NABL) Center have used hydraulic pump tests and field evaluations to create reference materials regarding oxidation studies for possible use with vegetable based lubricants. Below are examples of some of the test methods used to determine the oxidation stability of vegetable oils.

7.1.1 Active Oxygen Method (AOCS Method Cd 12-57)

In the active oxygen method (AOM), oxygen is bubbled through an oil or fat which is held at 36.6 °C (97.8 °F). Oil samples are withdrawn at regular intervals and the peroxide value (PV) is determined. The AOM is expressed in hours and is the length of time needed for the PV to reach a certain level. AOM is used as a specification for fats and oils. AOM hours tend to increase with the degree of saturation or hardness. This method has been largely replaced with Oil Stability Index.

7.1.2 Peroxide Value (AOCS Method 8b-90)

This is a test for measuring oxidation in fresh oils, and is highly sensitive to temperature. Peroxides are unstable radicals formed from triglycerides. A PV over 2 is an indicator that the product has a high rancidity potential and could fail on the shelf.

Figure 7.1 Oil stability instrument (top) and conductivity sensor in deionized water. See Plate 12 for the color figure

7.1.3 Oil Stability Instrument (AOCS Method Cd 1 2b-92)

Like AOM, in OSI, instead of pure oxygen, regular shop air is used and OSI is simpler to operate than AOM. A conductivity probe monitors conductivity of deionized water as evaporatives from test oil are emitted into the deionized water (Figure 7.1).

OSI values are expressed in hours. The lower the number of OSI hours, the lower the stability of the oil. For lubricants like hydraulic fluids that tend to reside in the system for hundreds or thousands of hours, a high OSI value for the base oil and yet a higher value for the formulated products will be needed. OSI values can be correlated with other oxidation tests. But, for vegetable oil based lubricants, it is best to establish a relationship between field test results and the OSI values. Table 7.2 shows the OSI values for selected vegetable oils [2]. The operation of this machine is similar to that of Rancimat as explained below and Figure 7.2 shows a schematic of the conductivity cell.

7.1.4 Rancimat

Rancimat [3] is a test that provides similar information to the results from the OSI test. It was recently approved by ASTM as a part of the Biodiesel standard test ASTM D 6751 (Figure 7.2).

Table 7.2 OSI values for selected vegetable oils

Oil	OSI (hours)
Apricot kernel	23.42
Avocado	18.53
Babassu	57.80
Castor	105.13
Coconut	75.38
Corn	3.73
Cottonseed	4.35
Flaxseed	1.17
Grapeseed	2.83
Hempseed	0.10
Jojoba – refined	42.15
Jojoba – golden	38.30
Macadamia	6.87
Olive	5.08
Palm	21.52
Poppyseed	17.86
Ricebran	20.82
Ricinoleic acid	117.10
Safflower	17.98
Sesame	5.80
Soy	17.67
Sunflower	10.23
Walnut	16.48

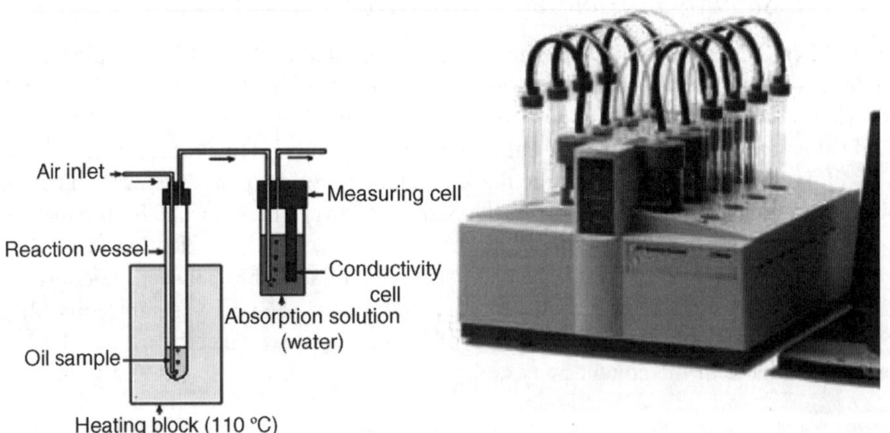

Figure 7.2 Components of Rancimat with oil sample and conductivity probe shown in deionized water [3]

7.1.5 *Viscosity Change as a Measure of Oxidation*

Honary [4] reported that using a hydraulic pump test would expose the vegetable oil to industrial conditions that are difficult to duplicate in many standard test methods. For example, he reported that most untreated vegetable oils may meet the ASTM wear protection requirements for hydraulic pumps because they naturally adhere better to metal surfaces, which prevents boundary lubrication but also, these oils break down and oxidize. In ASTM D 2882, currently designated as ASTM D 7043, the untreated crude soybean oil showed 40 mg wear, which is under the 50 mg passing maximum level. The lack of oxidation stability, however, resulted in *increased viscosity* of the oil, which in extreme cases could lead to polymerization.

As a result, in studying the performance of vegetable oil, vis-à-vis resistance to oxidation, some hydraulic pump tests could be used with the primary purpose of monitoring the changes in viscosity. Empirical research combined with field test observations indicate that an increase of less than 10% in viscosity for an oil tested in an ASTM D 2271 (currently designated as ASTM D 7043) or equivalent is desirable. It is also observed that, although the addition of additive packages to vegetable oils improves many of the oils' characteristics, this could negatively impact its natural lubricity.

For vegetable-based hydraulic fluid, long-term tests like ASTM D 4073 (formerly ASTM D 2271), a1000-hours test, are more desirable than shorter-term tests such as ASTM D 2882 (100 hours). Increased oleic acid content combined with a reduced percentage of linolenic acid, which is developed through new genes, could present improved oxidation stability as well as reducing the cost of the base oil by eliminating the need for chemical modifications such as hydrogenation.

Figure 7.3 shows the arrangement of the components of the hydraulic system set up according to the ASTM D 7043. The pump cartridge components are also shown separately, comprising two brass side plates, a rotor, and a cam ring and vanes. The cam ring and the vanes are weighed before and after the test and the weight loss is used as a determining factor for pass or fail. To pass the test, a weight loss of less than 50 mg would be required.

The speed for running the pump in this test is 1200 rpm, and the reservoir is stainless steel placed 60 cm (2 feet) from the inlet of the pump. A pressure relief valve allows pressure adjustment and a cooler maintains the temperature of the oil at test conditions. The pump has a flow rate of about 30.3 liters (8 gallons) per minute.

Table 7.3 shows the change in viscosity for a number of vegetable oils when exposed to the hydraulic pump test conditions requiring 1000 hours of exposure to 7 MPa (1000 psi) at 79 °C (174 °F). In order to show the relationships between fatty acid make up and stability, as shown by changes in the viscosity, the percent of oleic acid (18:1) is listed for each oil. Table 7.3 shows that a higher oleic acid content improves oxidation stability resulting in less of an increase in the viscosity of the oil in the test. In cases where the oleic acid contents are the same, the percent of other fatty acids, especially linoleic (18:2) and linolenic acid (18:3), was considered the reason for the higher change in viscosity.

An example of the viscosity increase in the ASTM D 7043 (formerly ASTM D 2271) is shown in Figure 7.4 and Table 7.4. Untreated high oleic soybean oil was tested and compared to the viscosity increase for conventional crude soybean oil. The viscosity increase for the commodity soybean oil was about 43.86 cSt or 146% with the high oleic soybean oil showing negligible change in viscosity at about 1.2%. The high oleic soybean oil had an oleic acid content of 83% and a linolenic acid content of 1.75%.

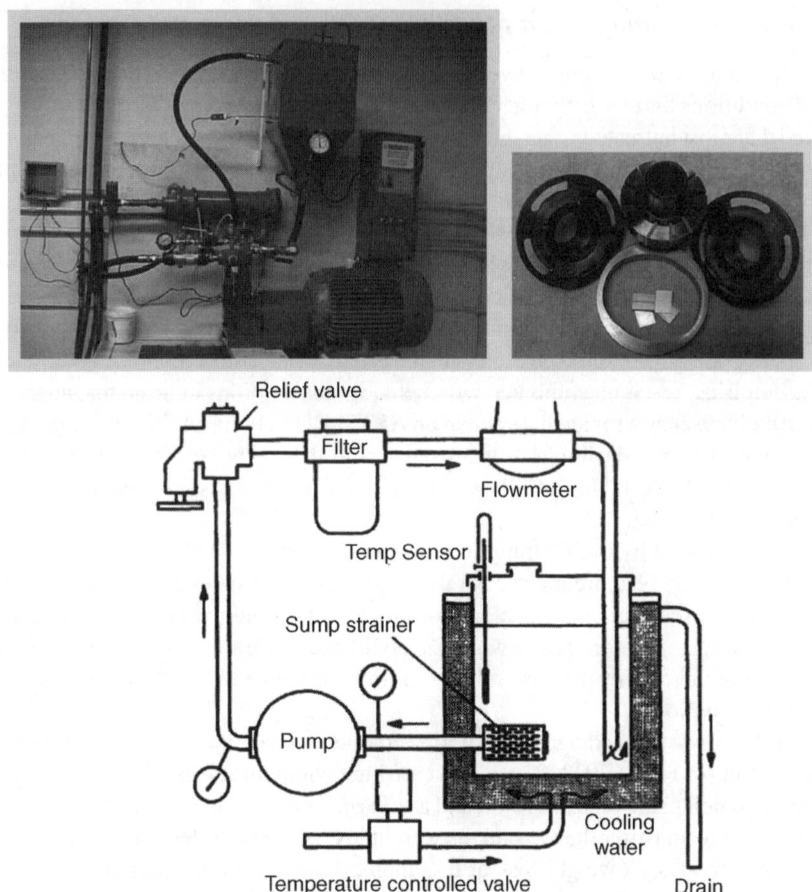

Figure 7.3 System set up for the ASTM D 7043 (formerly ASTM 2882 and ASTM D 2271) – actual hydraulic system, pump cartridge components and schematic. See Plate 13 for the color figure

Table 7.3 Changes in the viscosity of oils as tested in ASTM D 7043 (formerly ASTM D 2271), 1000-hour pump test

Oil type	% Oleic acid	Δ Viscosity (cSt)
Crude soybean oil (hexane extracted)	23.4	43.86
Low linolenic soybean oil	32.9	20.57
Partially hydrogenated soybean oil	37.5	24.18
Crude soybean oil (mechanically extracted)	23.4	19.15
High oleic sunflower oil	78.2	19.24
Ultra high oleic sunflower oil	86.8	16.23
High oleic canola oil (supplier 1)	76	14.08
High oleic canola oil (supplier 2)	76.5	19.53
Palm oil	38.8	12.97
Meadowfoam	62.5	32.55

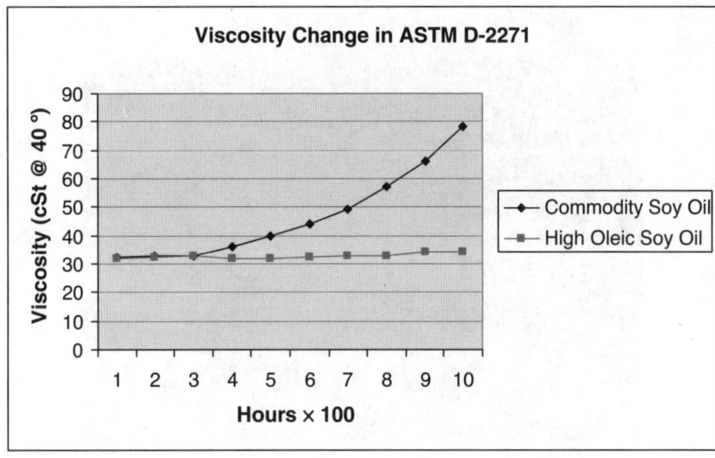

Figure 7.4 Changes in viscosity for commodity and high oleic soybean oils in ASTM D 7043 (Formerly ASTM D 2271)

7.2 Applications

The low stability oil can be used for applications where the lack of oxidation stability is desired. Linseed oil, for example, is used in artists' oil paints. An application, like a vegetable-oil based dust suppressant, is another example where lack of oxidation stability is desired. In this case the oil or a formulated product should remain a liquid and remain stable until it is applied to the area needing dust control (Figure 7.5). Afterwards, the oxidation or the polymerization of the oil might become an advantage if it forms a thin coating over the dusty area. Conversely, an application like hydraulic oil would require a much more stable oil because the oil would be in a system for a long time and be exposed to many possible catalysts that would increase oxidation.

Another example is the use of a vegetable oil for wood treatment. In a patented process [4] oxidized soybean oil is mixed with various preservatives and other performance enhancing chemicals to be forced into wood in a pressure treating chamber. After treating and subsequent drying of the outside of the wood, the oil inside the wood is oxidized and eventually cured into a polymer. Figure 7.6 shows treated wood specimens that are pressure treated with only oxidized

Table 7.4 Changes in the viscosity of formulated oils as tested in ASTM D 7043 (ASTM D 2271), 1000-hour pump test

Oil type	Δ Viscosity (cSt) %
Soybean oil-based tractor hydraulic fluid	4.60
Canola oil-based tractor hydraulic fluid	2.93
Rapeseed oil-based hydraulic fluid	37.20
Petroleum based hydraulic fluid	1.50
50-50 Blend of soybean oil-based tractor hydraulic fluid and petroleum-based hydraulic fluid	1.60

Figure 7.5 Manual application of soybean oil-based water emulsified dust suppressant

soybean oil for experimental railroad tie and with a mixture of oxidized soybean oil and copper naphthenate for home decking boards.

After curing, the pores of the wood are filled with polymerized oil, which creates a barrier to water. Figure 7.7 shows the inside of treated wood where the pores are filled solid with polymers.

Another application where low oxidation stability could be helpful is in rail curve grease. In this application, soybean oil-based grease was formulated and applied to the gage face of the rail for lubricity and friction reduction purposes. Using a less stable oil would allow the grease to retain its flowability for a period of time, and the thin layer of grease exposed to track metal

Figure 7.6 Pressure treating railroad tie with oxidized soybean oil (left); Decking boards with oxidized soybean oil and copper naphthenate. See Plate 14 for the color figure

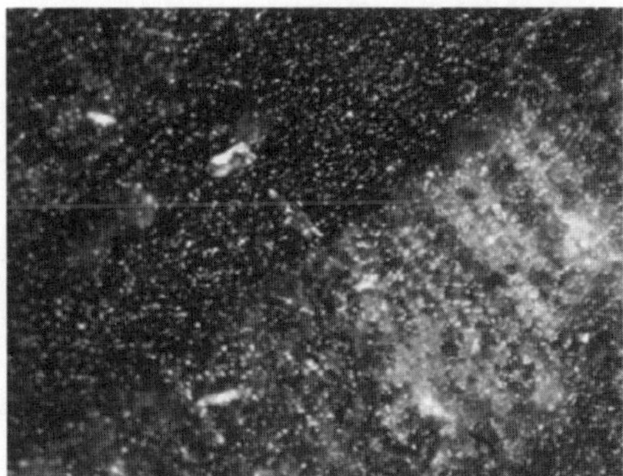

Figure 7.7 Cross-sectional cut of sample treated wood. See Plate 15 for the color figure

and environmental elements would cause it to form a thin layer of polymeric coating. This coating would fill the microcracks and any imperfection in the metal surface, simultaneously preventing metal-to-metal contact more effectively.

7.3 Petroleum White Oil and Food Grade Lubricants

Purified mineral oils are known to be benign in the human digestive system. As the oil is not digestible, it would pass though the digestive tract with various benefits. White oils are purified petroleum oils that are considered purified for use as a base oil for food-grade lubricant formulation. Food-grade lubricants include applications in food-processing plants, bakeries, consumer products, and in other such applications that require exceptionally pure oil. White oils meet or exceed the highest industry standards for purity and stability.

White oils are manufactured with a mixture of paraffinic and naphthenic hydrocarbons purified to the highest degree. The base white oils are available in grades: 50–60 SSU (ISO VG 7), 65–75 SSU (ISO VG 10/15), 80–90 SSU (ISO VG 15), 200–215 SSU (ISO VG 32/46), and 340–365 SSU (ISO VG 68) [5]. To improve the stability of the oil for lubricant or grease application, vitamin E is added as a natural oxidation inhibitor. These oils are colorless, odorless, and tasteless before use, although during service life they may become contaminated with wear metal particles. They are generally non-toxic and have outstanding storage stability. Researchers at the University of Northern Iowa's NABL Center have shown that neat white oil passes an aquatic toxicity test using *Daphnia magna* (OECD 202) in the same way as neat soybean oil.

When used in applications where direct, indirect, or incidental contact with food may occur, white oils meet FDA requirements. They may be used for lubricating machinery, to protect against moisture, or to control dust on grains and animal feed. It also is used as a component in certain consumer products, such as cosmetics and pharmaceuticals, and may be used as smoke oil at air shows.

Other applications of white oils include bottling and canning equipment requiring a National Sanitary Foundation (NSF) H1 rating (formerly rated by USDA), protective coating for raw fruits and vegetables, sealants for eggshells, drip oil for deep well water pumps, process oil or diluents in adhesives, sealants, caulks, pharmaceuticals, cosmetics, textile lubricants, household cleaners and polishes.

White oils meet the requirements of the FDA code of federal regulations including:

- 21 CFR 172.878 for direct food contact;
- 21 CFR 178.3620 for indirect food contact;
- 21 CFR 573.680 for animal feed; and NSF International and former 1998 USDA H1 guidelines for incidental food contact;
- CTFA (Cosmetics, Toiletries and Fragrances Association); and if needed kosher and halal certification can be attained [5].

Since these oils are used in food applications, in the United States, until 1998 USDA approval was required. Then USDA decided to allow the industry to manage the labeling and certification. The NSF now certifies products that are to be used in food processing operations. The NSF reviews the formulation of a product that is to be certified and checks the components of the products against a list of products that are not allowable for food contact. Additives used for lubricants may also have NSF certification, thus making it easier to evaluate the product if the additive is already approved. Vegetable oil-based lubricants as base oil can easily be substituted for the mineral white oil. If the additive components or packaging are food grade also, then the formulated product can be certified as food grade.

Because the food grade oil or grease may come in to contact with the food, there are restrictions on the type of additives that can be used in their formulations. For example, the presence of many metals, such as sulfur, commonly used as extreme pressure (EP) additives, may not be acceptable in food grade hydraulic oil. This would require the manufacturer of lubricants to search alternative EP additives, which could impact both the performance and the price.

Being aware of this limitation, manufacturers of equipment for food processing machinery design the equipment such that the formulation of necessary lubricants is simplified. Most

Table 7.5 Examples of typical properties of varied viscosity base white oils

White oil grade	Typical properties				
	50-60	65-75	80-90	200-215	340-365
ISO grade	7	10-15	15	32/46	68
Specific gravity @ 15.6 °C (60 °F)	0.827	0.85	0.855	0.862	0.875
Density (kg/l) (lb/gal)	0.83 (6.89)	0.85 (7.09)	0.86 (7.13)	0.87 (7.19)	0.88 (7.3)
Gravity (°API)	39.6	35	34.1	32.7	30.2
Flash point (COC), °C (°F)	153 (307)	177 (351)	186 (367)	210 (410)	225 (437)
Pour point, °C (°F)	−9 (16)	−15 (5)	−15 (5)	−15 (5)	−15 (5)
Color, ASTM D1500	L0.5	L0.5	L0.5	L0.5	L0.5
Color, Saybolt	+30	+30	+30	+30	+30
Viscosity (cSt @ 40 °C)	6.9	12.3	15.2	39.5	66.2
Viscosity (cSt @ 100 °C)	2.4	2.9	3.4	6.1	8
Viscosity index	89	68	88	98	84

hydraulic pumps, for example, are derated to work with lower pressures if they are to be used for food processing applications. As a result, a food grade hydraulic oil may not require the same amount of antiwear or extreme pressure additives as is required for the performance level of industrial hydraulic oils.

Table 7.5 shows properties of commercial white oils from one supplier indicating the typical parameters for various viscosity grades.

Vegetable oils naturally portray performance attributes that are comparable to the same as those possessed by white mineral oils. As a result, provided they have adequate oxidation stability, they can be formulated with food grade additives to perform in food processing applications. Regardless of whether the base oil is white mineral oil or vegetable oil, caution needs to be exercised in their use.

References

1. Honary, L.A.T. (1995) Performance of Selected Vegetable Oils in ASTM Hydraulic Tests. SAE Technical Papers, Paper 952075.
2. Honary, L.A. (2002) Internal data from research performed at the University of Northern Iowa, National Ag-Based Lubricants Center.
3. Technical data from Brinkmann Company, retrieved April 18, 2010. http://www.brinkmanncanada.com/products/ppm_rancimat873_de.asp.
4. Honary, L.A.T. U.S. Patent # 6641927 "Soybean Oil Impregnation Wood Preservative Process and Products" issued on November 3, 2003.
5. Technical data from ConocoPhillips Company (2006). Retrieved April 18, 2010 from: http://www.seversonoil.com/pdfs/FamilyOfBrands/FO_White_Oil.pdf.

8

Performance Properties of Industrial Lubricants

8.1 Introduction

The development of biobased lubricants and greases requires that the physical and chemical properties of the starting materials are documented. This information is a requirement because it determines those modifications that are necessary to produce a lubricant suitable for its intended use. Such information may include the following:

1. Viscosity and viscosity index
2. Flash and fire points
3. Cold temperature performance
   a. Pour point
   b. Brookfield viscosity performance
   c. Storage at low temperatures
4. Total acid number
5. Iodine value
6. Water content
7. Rust and corrosion resistance
   d. Humidity cabinet tests
   e. Copper corrosion
8. Foaming characteristics
9. Fluid compatibility
   f. Compatibility with other fluids
   g. Elastomeric compatibility
   h. Compatibility with metallic components
   i. Homogeneity and stability of mixture
   j. Cleanliness.

Biobased Lubricants and Greases: Technology and Products, First Edition. Lou A.T. Honary and Erwin Richter

10. Demulsibility

Additionally, the oxidation stability of the oil would need to be evaluated using either:

11. Oil stability index; and/or
12. Viscosity stability in long run pump tests as described earlier. After determining these properties, additives are used to enhance the performance using friction, wear, and extreme pressure testing instruments including:
    a. Extreme pressure performance
    b. Timken OK Load Test
    c. Four Ball Extreme Pressure Test
    d. Pin and Vee Block Test (film thickness performance)
    e. FZG Test (Fur Zahnrader und Getriebebau (FZG) Test).
13. Antiwear performance including:
    a. Four Ball Wear Test
    b. Pump tests

    Finally, once the products are formulated, an evaluation of the environmental performance of the final product will be conducted using any or all of the following test methods:
14. Biodegradability – test methods ASTM D 5864 or OECD 301-A through F
15. Plant toxicity – test method OECD 308
16. Aquatic toxicity – test methods ASTM D 6081 or OECD 202 and OECD 203
17. Percent biobased content – test method ASTM D 6866
18. Life cycle analysis according to methods developed by the National Institute of Standards and Technology (NIST).

8.2 Common Performance Requirements

8.2.1 Viscosity

Defined as the (moving) fluid's resistance to flow, viscosity is normally tested using the Kinematic Viscosity method based on ASTM D 445. This tests uses velocity, or the time it takes for a certain volume of oil at a specified temperature to pass through a standard capillary tube. Its unit of measurement is often expressed in centi-Stokes (cSt) or mm^2/s. Generally, the kinematic viscosity is tested at 40 °C (104 °F) and at 100 °C (212 °F). The viscosity index (VI), which is *an indicator of the oil's resistance to change as temperature changes*, is then calculated using the values obtained at 40 °C and 100 °C, based on ASTM D 2270. Vegetable oils have a higher VI than equivalent viscosity petroleum oils with the same viscosity. For example, soybean oil having a kinematic viscosity of 32 cSt at 40 °C has a VI of 220, as compared to a mineral oil with the same viscosity having a VI of 95. This makes the viscosity of the vegetable oil more stable as the temperatures changes. Similarly, paraffinic oils have higher VI than naphthenic oils.

Figure 8.1 shows an automatic viscosity tester with the oil sample cups for oils to be tested at 40 °C and 100 °C. This test machine determines the viscosities and then calculates and reports the VI. Figure 8.2 shows a manual viscosity tester and associated glassware with standardized orifices.

Measuring viscosity at other temperatures would require different calculations. For temperatures over 100 °C, an oil is used instead of water in the bath because water boils at this temperature.

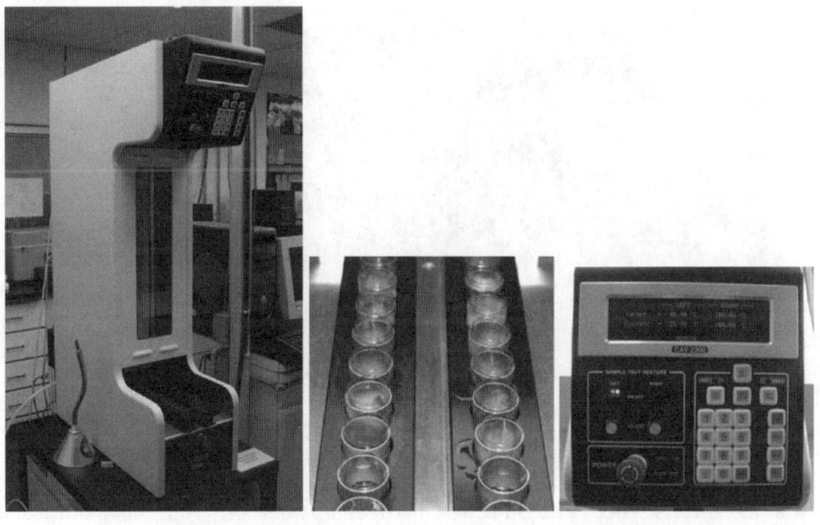

Figure 8.1 Automatic viscosity tester (left) with cups for viscosity at 40 °C and 100 °C

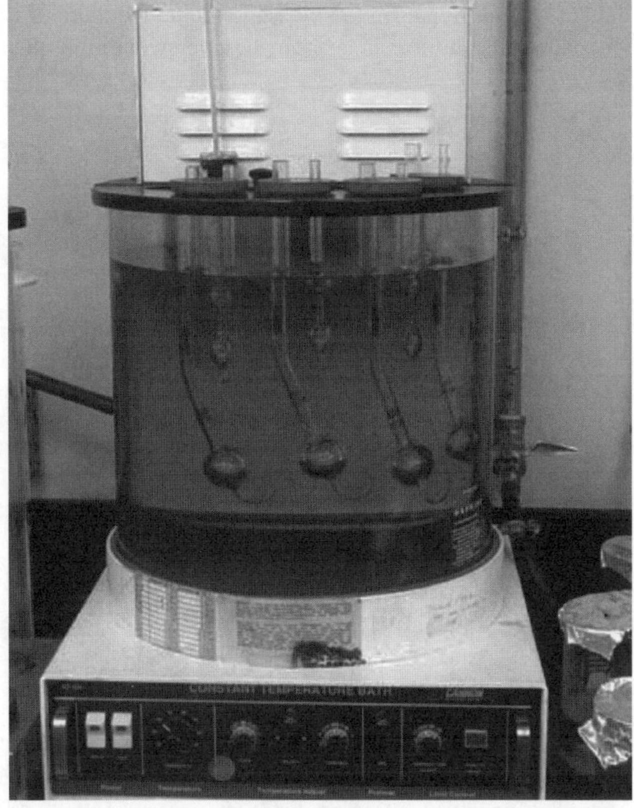

Figure 8.2 Manual viscosity tester. See Plate 16 for the color figure

Figure 8.3 Cold temperature viscosity tester

Similarly, to test the viscosity of an oil below the freezing point of water, fluids other than water need to be used. Figure 8.3 shows an automated cold temperature viscosity tester.

For lubricants designed for outdoor use, like in tractor hydraulic and transmission fluids, the Brookfield viscosity test is often specified. The Brookfield viscosity tester has a rotating shaft with a standard dimension that rotates in the oil while the temperature of the oil is reduced (Figure 8.4). This, of course, results in an increase in the amount of torque on the shaft. This test

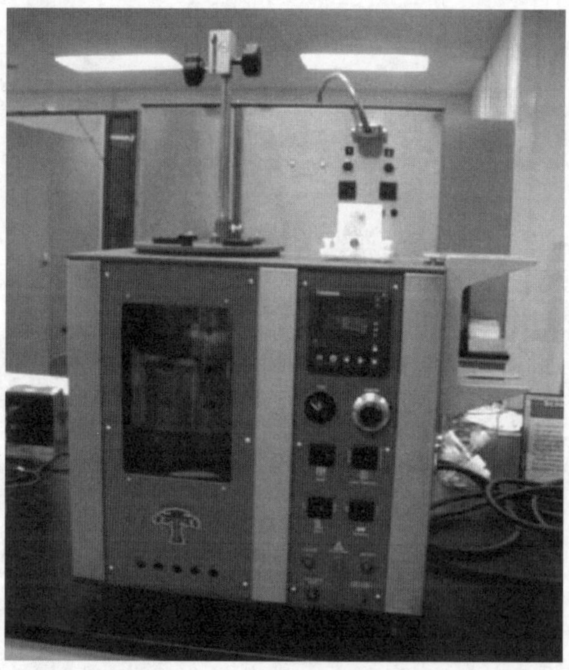

Figure 8.4 Brookfield viscosity tester. See Plate 17 for the color figure

Plate 1 Figure 1.10 Old's curved dash vehicle started the mass assembly trend. Picture of Eli Olds Car advertisement

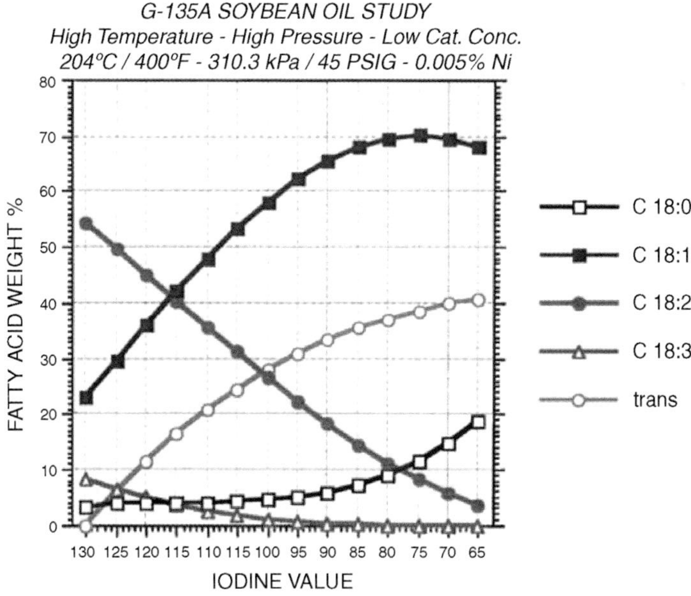

Plate 2 Figure 4.6 Formation of fatty acid in hydrogenation process

Plate 3 Figure 4.7 Soybean field (left) and Soybeans ready for harvest

Plate 4 Figure 4.13 Corn

Plate 5 Figure 6.2 Guayule field (left) and natural guayule rubber (right)

Plate 6 Figure 6.3 Examples of industrial crops for biobased fuels and lubricants

Plate 7 Figure 6.6 Cuphea

Plate 8 Figure 6.7 Castor plant. Courtesy of the Schundler company

Plate 9 Figure 6.9 Jatropha plant. Photo courtesy of the Centre for Jatropha Promotion and Biodiesel India

Plate 10 Figure 6.12 Poppy seed [23] and poppy field

Plate 11 Figure 6.14 (a) Coconut tree, (b) coconuts and coconut oil, and (c) illustration of coconuts [23]. Reproduced from Mt Banahaw Health Products website www.coconutoil.com

Plate 12 Figure 7.1 Oil stability instrument (top) and conductivity sensor in deionized water

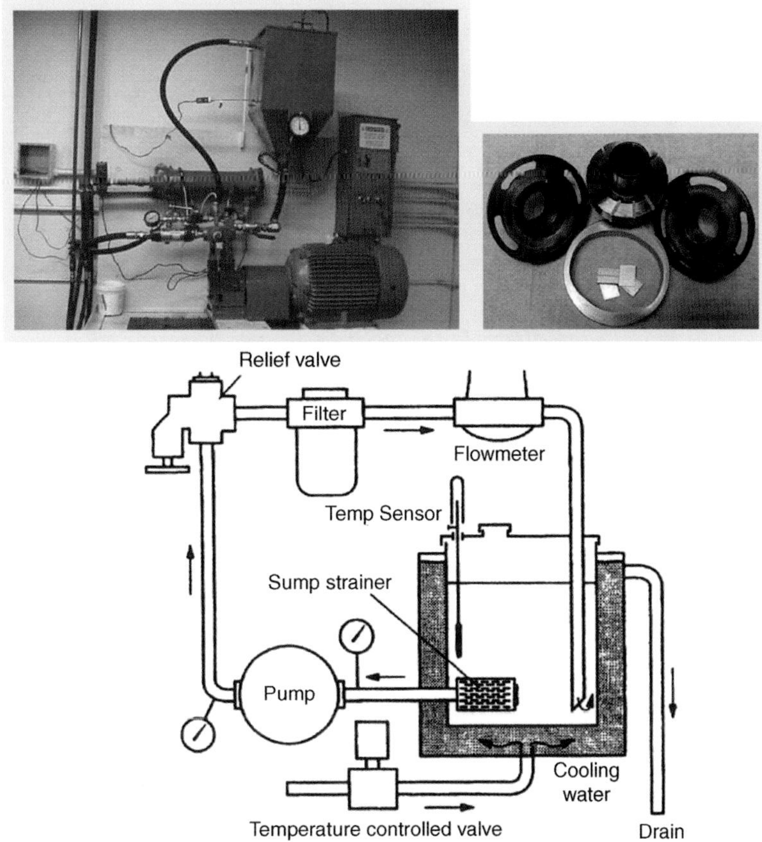

Plate 13 Figure 7.3 System set up for the ASTM D 7043 (formerly ASTM 2882 and ASTM D 2271) – actual hydraulic system, pump cartridge components and schematic

Plate 14 Figure 7.6 Pressure treating railroad tie with oxidized soybean oil (left); Decking boards with oxidized soybean oil and copper naphthenate

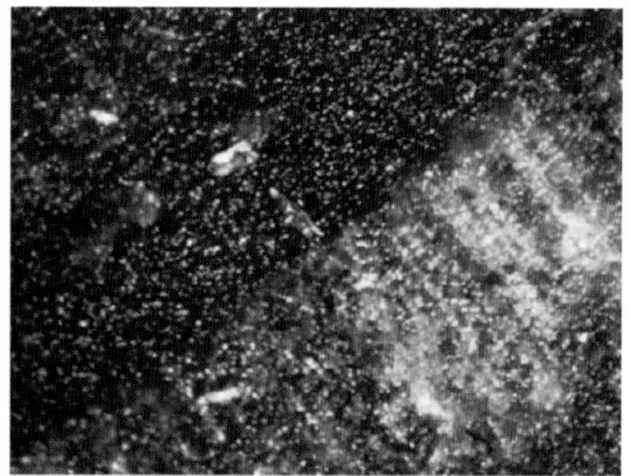

Plate 15 Figure 7.7 Cross-sectional cut of sample treated wood

Plate 16 Figure 8.2 Manual viscosity tester

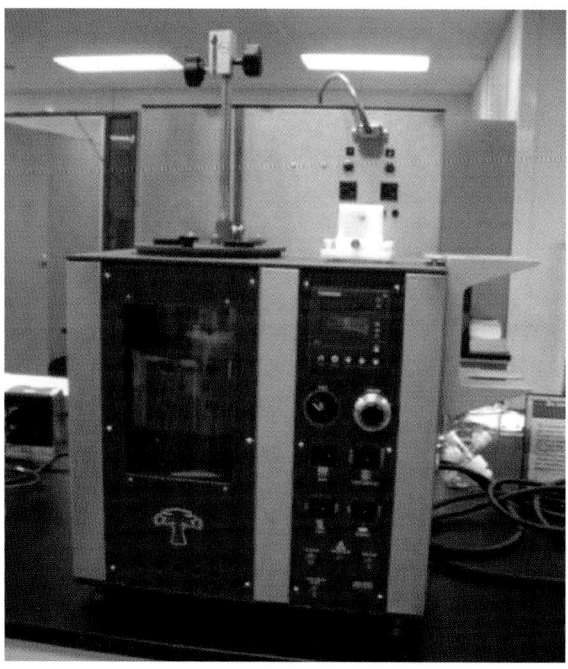

Plate 17 Figure 8.4 Brookfield viscosity tester

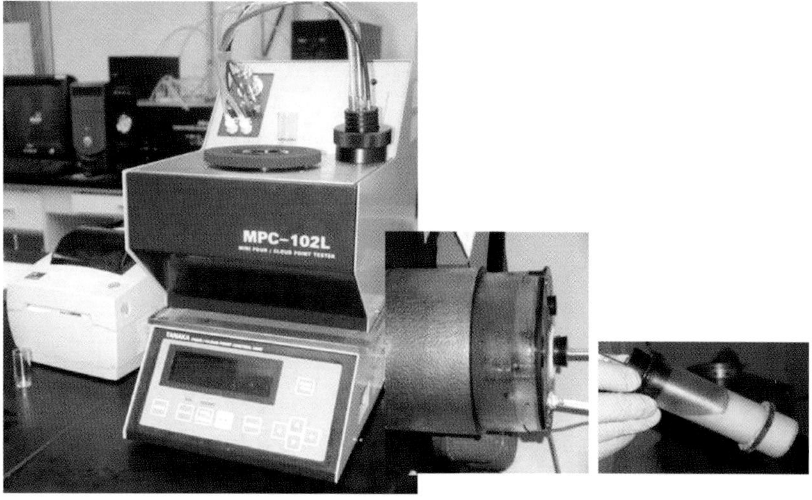

Plate 18 Figure 8.7 Automatic pour point tester (left) and manual unit

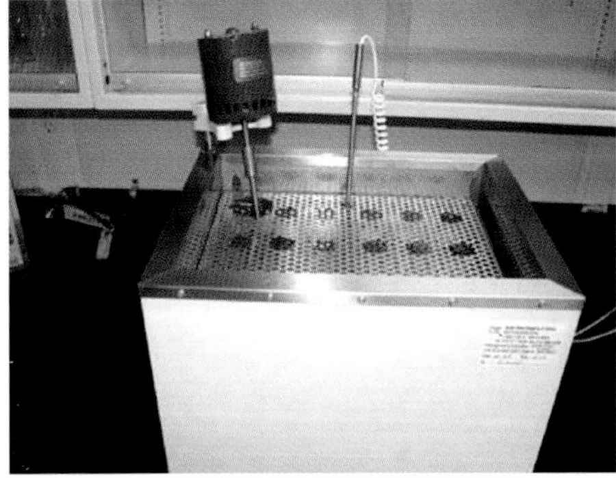

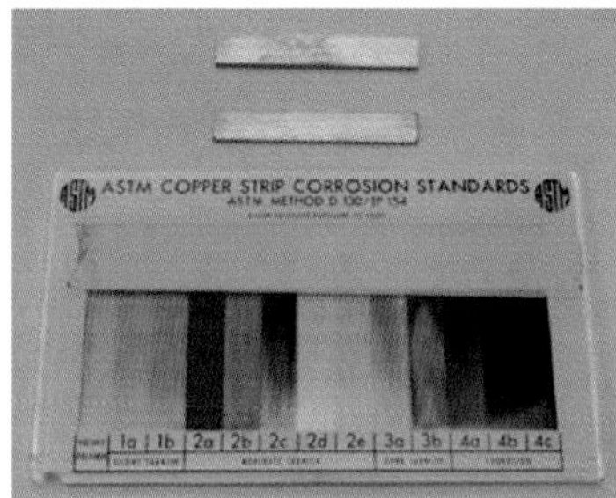

Plate 19 Figure 8.9 Copper corrosion tester (left) specimens and ASTM Strip Standards

Plate 20 Figure 8.14 Dielectric fluid tester

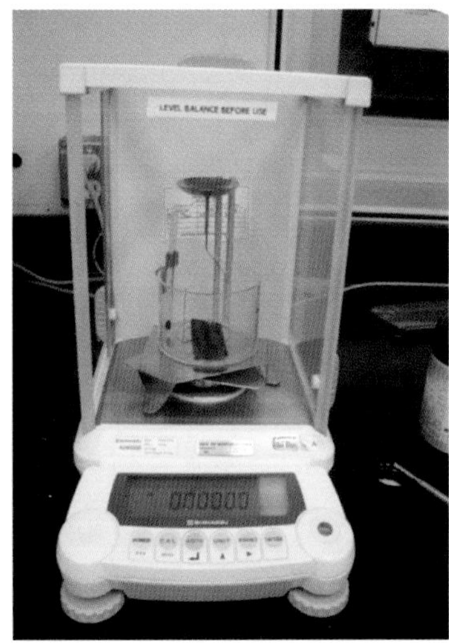

Plate 21 Figure 8.16 Measuring weight change due to elastomer swelling

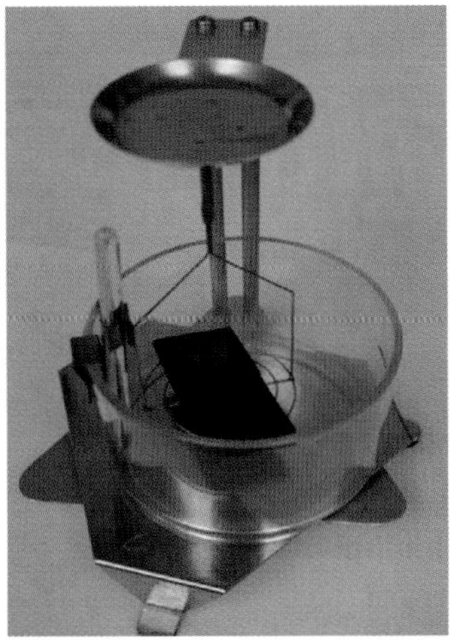

Plate 22 Figure 8.18 Special attachment for weighing rubber sample

Plate 23 Figure 8.21 Rotary Bomb Oxidation Test (RBOT) fresh and spent copper catalysts

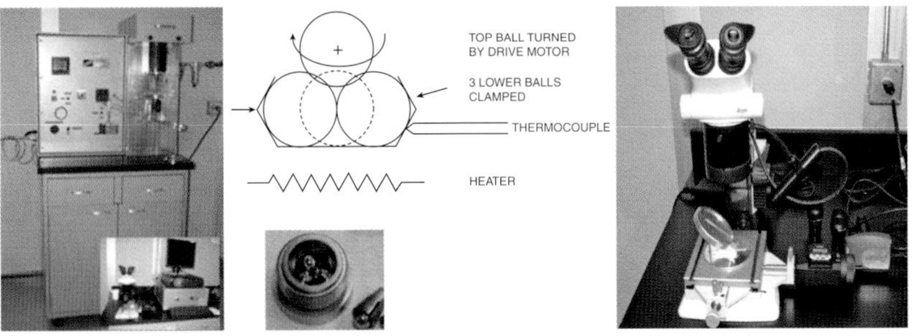

Plate 24 Figure 8.26 Four Ball Wear Tester (left) with microscope for scar diameter measurement

Plate 25 Figure 8.30 Test tubes containing soybean oil-based straight metalworking fluids and water emulsified coolants

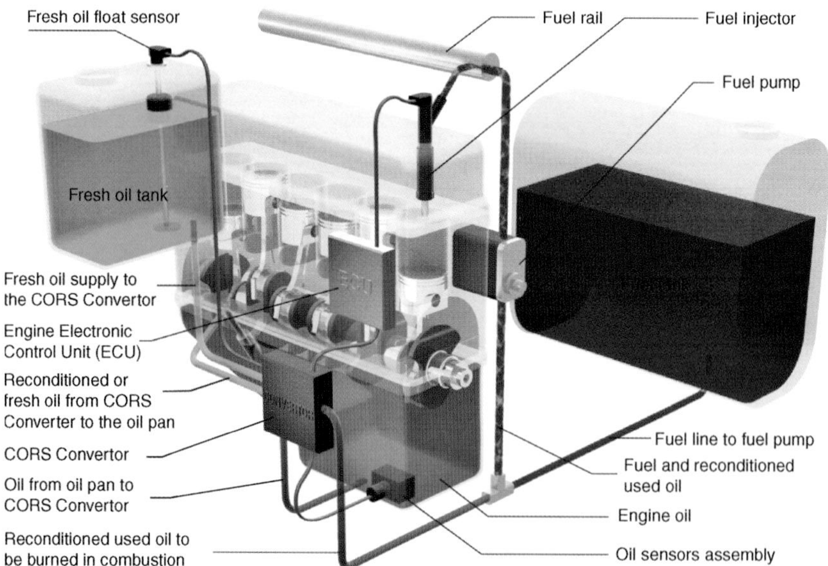

Plate 26 Figure 8.37 Conceptual representation of the CORS components plus the engine ECU and injector

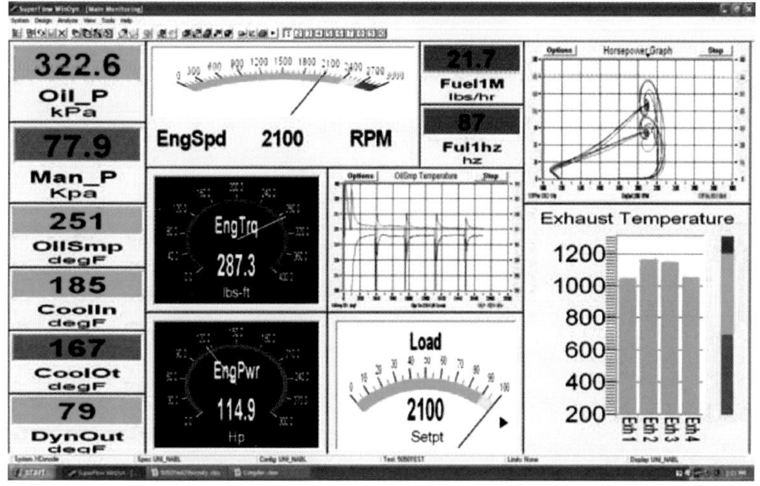

Plate 27 Figure 8.39 A screen shot of the parameters set for the engine tests

Plate 28 Figure 9.7 Dehydrated preformed soap (left) and resulting grease (right)

Plate 29 Figure 9.14 Microwave grease reactor in back with two 75 KW transmitters in front and waveguides for transferring microwave energy to the reactor

Plate 30 Figure 9.26 Wayside automated rail curve grease lubricator

Plate 31 Figure 9.28 Operation of hand-held tribometer

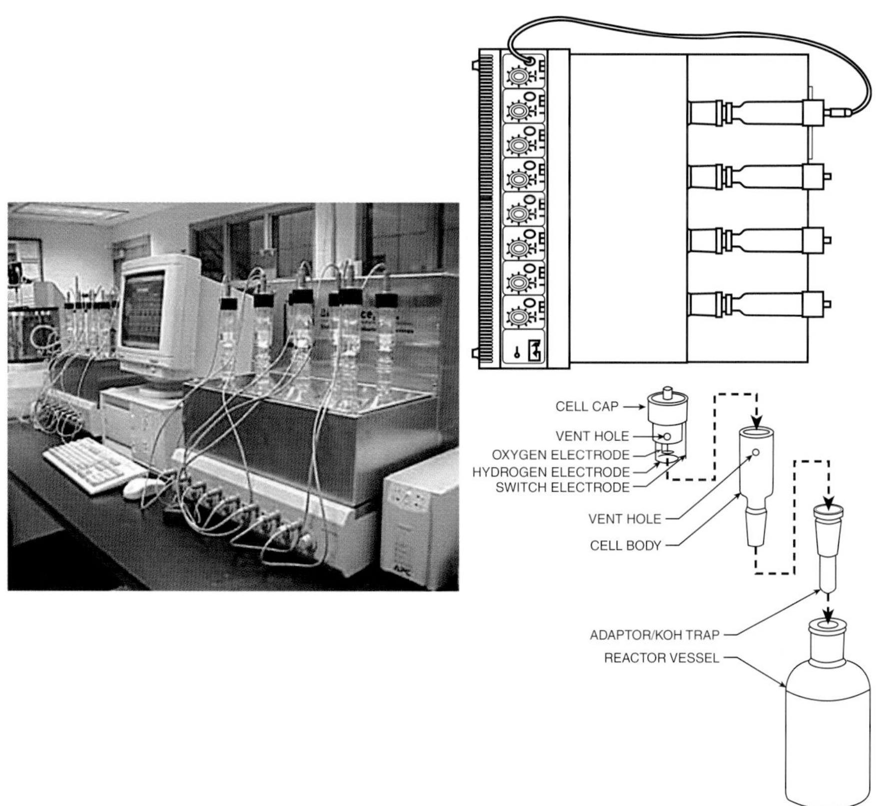

Plate 32 Figure 10.2 Electrolytic respirometer – and a view of a standards reactor assembly (right bottom)

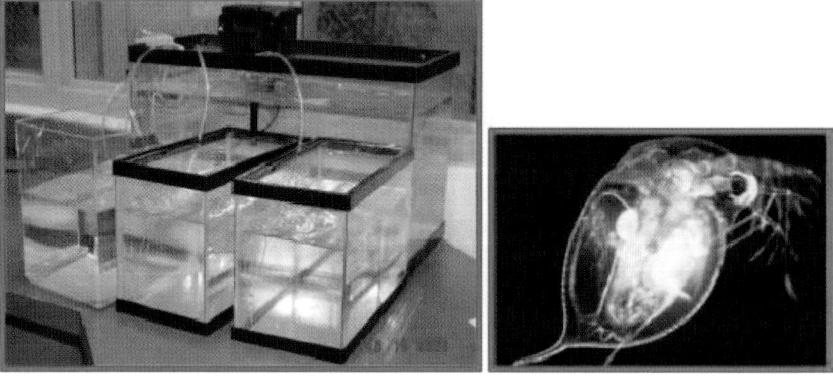

Plate 33 Figure 10.3 *Daphnia magna* can be grown and fed in conventional fish tanks (left) and actual photo of Daphnia (right)

Plate 34 Figure 10.4 Plant growth chamber

is highly recommended for biobased hydraulic oils or oils that are expected to perform in the outdoors at cold temperatures. As the oil is cooled, the presence of any crystallized fat could cause the oil to behave unpredictably and thicken up. When using a scanning Brookfield viscosity test, the changes in the viscosity of the oil could be monitored for unusual peaks.

Dynamic viscosity is defined as a measure of a (static) liquid's resistance to movement. Dynamic viscosity is measured in units of Poise, often expressed in centiPoise (cP) or in pascal seconds (Pa s).

8.2.2 Flash and Fire Points

Using ASTM D 92 and 93, the flash and fire points of vegetable oils are tested in the same manner as petroleum products. Vegetable oils in general have flash and fire points of about 93.3 °C (200 °F), higher than equivalent viscosity petroleum oils.

Both open-cup flash point testers and closed-cup flash point testers are used. In a closed-cup flash/fire point tester (ASTM D 93), the oil to be tested is placed inside a standard cup which is closed by covering while it is heated. At regular intervals, as the temperature of the oil increases, substances with lower boiling points evaporate and burn when exposed to the open flame. The point that the oil evaporatives flash and burn is considered the flash point. In a closed-cup, the evaporatives remain under the cup cover and this could lead to a more accurate flash point indication. When the test fluid is heated to its fire point, then exposed to the open flame, it bursts into flame and sustains the fire (Figure 8.5).

In the open-cup test the same process is performed, but as the oil is heated any gas or evaporative material may escape from the cup because it is open. This may yield a higher flash point. The test is based on the ASTM D 92 method (Figure 8.6).

8.2.3 Boiling Range

Since oils are not uniform in makeup and are made of different molecules with their own boiling points, a boiling range is established encompassing the boiling points of the various

Figure 8.5 Closed-cup flash/fire points tester

Figure 8.6 Open-cup flash/fire points tester

light or heavy constituents of the oil. There is a relationship between boiling range, molecular weight, and viscosity. A higher molecular weight results in a higher viscosity and a higher boiling range.

8.2.4 Volatility

Volatility is associated with the flash point and is determined using the ASTM D 972 which is designed to measure the loss in mass of a substance after 22 hours of exposure to a standard temperature of 107 °C (225 °F). For lubricant applications, like metalworking fluids, a lower volatility is desired to avoid generating mist and to reduce fire hazards. Vegetable oils in general are less volatile than petroleum oils of the same viscosity.

8.2.5 Cold Temperature Properties

These properties are expressed in the form of cloud point and pour point. The pour point is measured using the ASTM D 97 method, which is the temperature at which the oil stops flowing (Figure 8.7). The cloud point is measured by determining the temperature at which fats in vegetable oils or waxes in mineral oils crystallize. At the cloud point, the oil no longer follows the principles of Newtonian fluids and becomes a multi-phase system of liquid and various crystallized or semicrystallized matters. Naphthenic oils have lower pour points than paraffinic oils. Vegetable oils containing unsaturated fatty acids have lower pour points than those with a higher content of saturated fats. Table 8.1 [1] shows the pour point of naphthenic and paraffinic oils.

Experience with formulated vegetable oil-based lubricants has shown that the pour point alone should not be relied upon as a measure of performance in cold temperatures. This is somewhat unique to vegetable oil-based technology because although a vegetable oil may have a pour point at a given temperature, that oil may still stop flowing at higher than its pour point

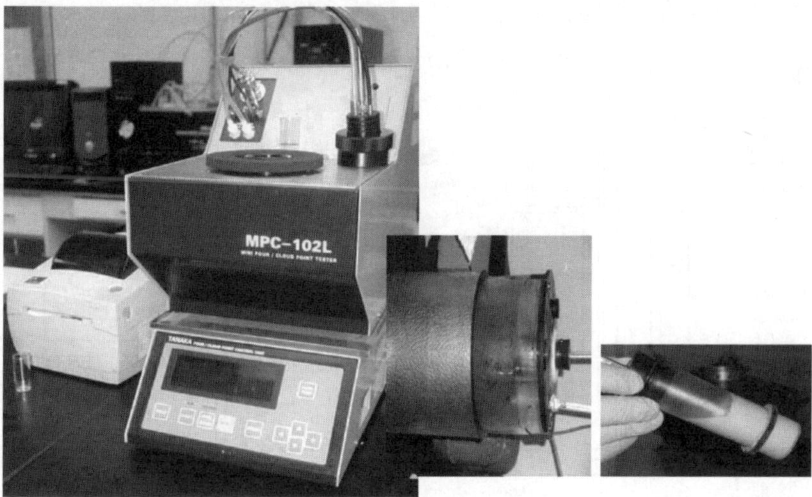

Figure 8.7 Automatic pour point tester (left) and manual unit. See Plate 18 for the color figure

when exposed to the cold over extended periods of time. As an example, a vegetable-based hydraulic oil having a pour point of −40 °C (−40 °F) may freeze at −28 °C if exposed to the −28 °C temperature for an extended period of time. This is due to the presence of different fatty acids which may begin to seed crystals that would in time solidify the entire body of the oil. Hence, for cold temperature applications, in addition to the pour point, long-term cold-storage and scanning Brookfield viscosity tests should be considered.

8.2.6 Density

Measured in g/cm^3, density is a measure of the mass of a material in a unit volume at 15 °C (59 °F). Measured according to the ASTM D 4502, density is affected by temperature. Density and temperature are inversely related.

8.2.7 Foaming Properties

These are tested using ASTM D 892. Typically, air is bubbled through a sample of oil, which is at a constant temperature in a bath, allowing the foam column to rise (Figure 8.8). After removing the air-flow, the time it takes for the foam column to collapse is recorded. Fluids with good antifoam additives do not sustain the foam and the foam collapses quickly. The ASTM D

Table 8.1 Pour points of naphthenic and paraffinic oils

Viscosity in cSt at 40 °C	Naphthenic pour point (°C (°F))	Paraffinic pour point (°C (°F))
30	−39 (−37)	−39 (−38)
110	−24 (−11)	−3 (27)
400-500	−18 (0)	−5 (23)

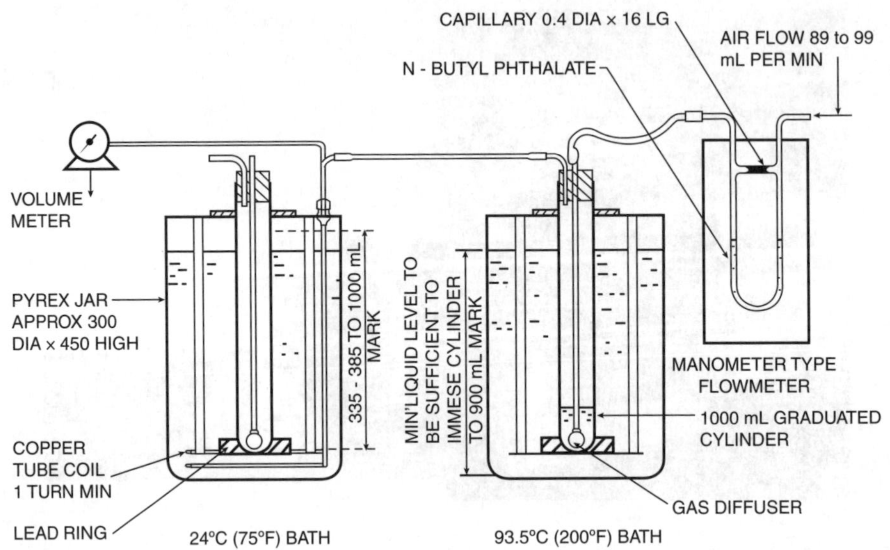

Figure 8.8 Schematic of foaming apparatus with two temperature baths for 25 and one for 95 °C. Reprinted, with permission, from D6082 Standard Test Method for High Temperature Foaming Characteristics of Lubricating Oils, Copyright ASTM International, 100 Barr Harbor Drive, West Conshohocken, PA 19428. A copy of the complete standard may be obtained from ASTM International, www.astm.org

892 test method covers the determination of the foaming characteristics of lubricating oils or other liquids at 24 °C. Means of empirically rating the foaming tendency and the stability of the foam are described. The inlet air pressure is set to 100 kPa (15 psi) to blow air through stone diffusers at 94 ± 5 ml/min. Desired sequence of tests are performed to ensure foaming is determined at different temperatures.

1. Sequence I
   a. Approximately 200 ml of sample oil is decanted into a beaker and heated to 49 ± 3 °C (120 °F), then allowed to cool to 24 ± 3 °C (75 °F).
   b. Within 3 hours of preparation:
      i. 190 ml of sample is poured into a 1000 ml cylinder
      ii. The cylinder is immersed to at least the 900 ml mark in the 24 ± 0.5 °C (75 °F) bath and the temperature allowed to stabilize
      iii. The gas diffuser and air inlet tube are inserted and allowed to soak for 5 minutes (with the air source not open yet!)
      iv. The air source is then connected and adjusted to the flow rate to run the air through the sample for 5 min $\pm$ 3s
      v. At this point the air source is turned off and the volume of the foam is recorded (this is the foaming tendency)
      vi. Again the sample is allowed to stand for 10 min $\pm$ 10 s and the recording the foam volume (this is the foam stability).

The results are reported in terms of Sequence I or II or III depending on the number of tests run at selected temperatures. Reporting includes foaming tendency, foaming stability, and foaming volume at the end of a 5-minute air flow and foaming volume at the end of a 10-minute settling period.

8.2.8 Copper Strip Corrosion

The fluid is tested with the ASTM D 130 Copper Corrosion Test, which is rated by the changes in the color of the test specimen, using reference color charts. The test procedure appears in the following section.

8.2.9 Copper Strip Corrosion Test

This test determines the degree of a lubricating oil's corrosiveness to copper. The operation of the tester can be summarized as follows: The oil bath is heated to 100 °C, a sample cup is filled with a sufficient amount of oil making sure there are no air bubbles, all six sides of a new copper test strip are sanded using silicon carbide paper until all blemishes are removed and then they are finished by sanding with a 240 grit silicon carbide paper. The copper test strip is washed in *n*-heptane to remove all dust and dirt, and it is polished again on all sides with 150 mesh silicon carbide grains picked up with cotton wool moistened with solvent until the entire surface of the strip is polished uniformly. The strip is wiped with the cotton wool to remove all dust and again rinsed with the solvent until clean (Figure 8.9).

1. The strip is placed in the middle of the sample cup and is completely covered by at least 5 mm of oil and then the tube is covered loosely with a watch glass.
2. The test is run for 3 hours.
3. The test strips are compared with the standards to determine the level of corrosion.

8.2.10 Rust Prevention

This is determined using ASTM D 1748, where humidity cabinets (Figure 8.10) are employed to determine the rust protection of the sample oil in high humidity.

The test procedure can be simplified in section 8.2.11.

Figure 8.9 Copper corrosion tester (left) specimens and ASTM Strip Standards. See Plate 19 for the color figure

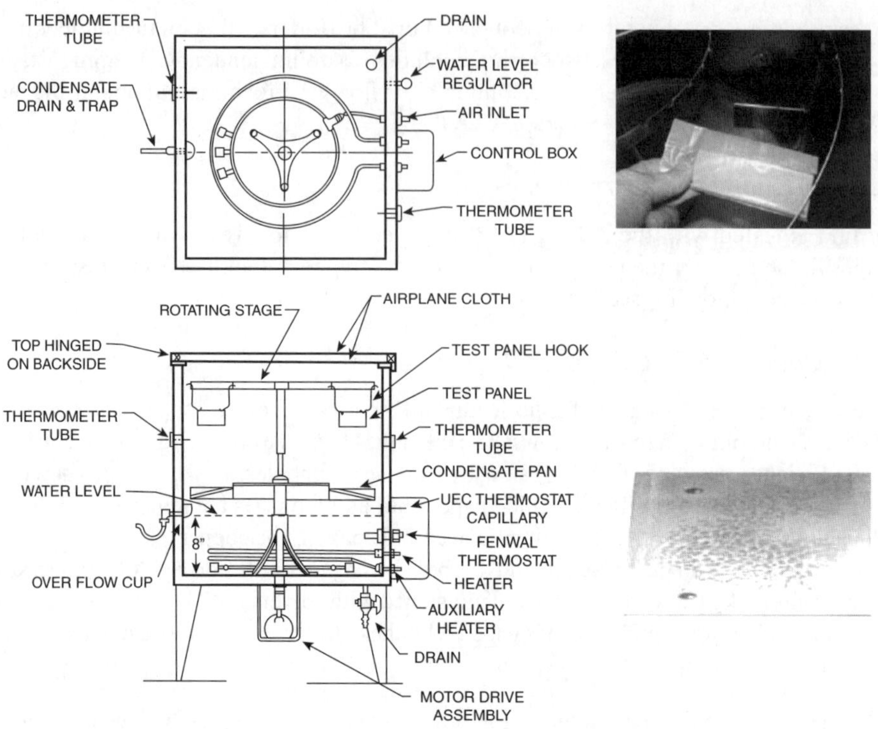

Figure 8.10 Schematic illustration of humidity cabinet (ASTM D 665) (left) and new (top) and failed (bottom) specimens

8.2.11 Test Purpose

For evaluating the rust-preventive properties of metal preservatives under conditions of high humidity based on ASTM D1748.

The dimensions of the standard are $51 \times 102 \times 3.2$ mm ($2 \times 4 \times 1/3$ in); two holes of 3.2–2.3 mm in diameter in the corners of the 102 mm edge and should weigh 110 ± 15 g. These panels come wrapped in plastic protective covers; keep the panel on a clean dry surface. Use a 240-grit aluminum oxide abrasive cloth; any rust or protective oil coating is washed away with petroleum naphtha and a lint-free cloth. Panels should be used the same day they are prepared.

The sample oil to be tested for rust preventative properties is prepared in a clean 400 ml beaker at a temperature of 23.3 ± 0.5 °C (74 °F). Each panel is removed from the methanol with a clean suspension hook. The panels are hung above boiling ASTM precipitation naphtha for 5 minutes making sure that the panel is completely wet with refluxing solvent and then allowed to air dry for 10–20 seconds. The panel is then placed into the sample oil for 10 seconds and withdrawn, then replaced into the oil sample again for 1 minute, and then removed slowly. At a temperature of 24.1 ± 3 °C (75 °F), for $2\,h \pm 20$ min, the panels are suspended in the humidity cabinet. The air temperature, pH, air rate, and water level are checked and adjusted at

7- to 8-hour intervals and the panels are checked twice daily. They are examined within 10 min after removal under a fluorescent light. The panel will pass if it contains no more than three dots of rust, no one dot larger than 1 mm in diameter. The panel fails if it contains one or more dots larger than 1 mm or if it contains four or more dots of any size. The reporting is based on the number of hours the panel was in the humidity cabinet, the number of test surfaces (two for each panel), and the number of passing test surfaces.

Many vegetable oil-based lubricants have shown good performance in this test. Due to the polarity of vegetable oils, the protective coating provided seems to persist. Humidity cabinet hours of 100 or more may be specified for different applications.

8.2.12 Neutralization Number

Using ASTM D 664, the neutralization of vegetable oils is determined.

8.2.13 Solubility

Solubility is an important property of base oils because often additives are used to enhance the natural properties of the base oil and to formulate products. A combination of the aniline point and viscosity gravity constant is used to determine the solubility of oils. The viscosity gravity constant (VGC) is determined using the ASTM D 2501, which is a calculated, dimensionless value. This is an indication of the solvency property of the fluid.

8.2.14 Aniline Point

This is another property related to the solubility of the oil. Using the ASTM D 611 method, the aniline point is determined as the lowest temperature at which an oil is fully miscible with an equal amount of aniline. The chemical formula of aniline is $C_6H_5NH_2$. Analine is a derivative of benzene and is a colorless oily liquid. The lower the aniline point, the better the solubility for that oil. In the ASTM D 611, it is the temperature at which the aniline dissolves in the oil. Using a U tube, 10 ml of the sample oil and 10 ml of the aniline are heated until the aniline dissolves in the oil. It is then cooled and the temperature monitored until the aniline separates from the oil. This temperature is reported as the aniline point. The lower the aniline point, the higher the solubility of the oil.

8.3 Heat Transfer Properties

Vegetable oils are used as a substitute for petroleum as quench oil or for cooling oil in a transformer. Factors such as viscosity and fatty acid makeup affect the heat transfer properties of the vegetable oil. Facina and Colley [2] reported on the viscosity and specific heat of vegetable oils as a function of temperature. The viscosities of 12 vegetable oils at temperatures from 35 °C (95 °F) to 180 °C (356 °F) are presented in Table 8.2. The viscosities at 35 °C were about 10- to 15-fold larger than viscosities 180 °C while the percent increase in specific heat from 35 to 180 °C was about 17% ([2], p. 741).

When quenching (heat treating) heated metals, the rate of cooling is important. Too fast a cooling rate could result in micro-cracks in the surface of the heat treated part. The depth of hardness can also be impacted. Cooled too slowly, the metallurgical properties could be impacted and the hardness level reduced. Vegetable oils due to their high flash and fire points

Table 8.2 Viscosities of 12 vegetable oils at various temperatures

Oil source	Sample temperature (°C)									
	35	50	65	80	95	110	120	140	160	180
Almond	43.98	26.89	17.62	12.42	9.15	7.51	6.54	5.01	4.02	3.62
Canola	42.49	25.79	17.21	12.14	9.01	7.77	6.62	5.01	4.29	4.65
Corn	37.92	23.26	15.61	10.98	8.56	6.83	6.21	4.95	3.96	3.33
Grape seed	41.46	25.27	16.87	11.98	9.00	10.37	9.18	7.50	6.10	4.78
Hazelnut	45.55	27.40	17.83	12.49	9.23	7.56	6.69	5.25	4.12	3.48
Olive	46.29	27.18	18.07	12.57	9.45	7.43	6.49	5.29	4.13	3.44
Peanut	45.59	27.45	17.93	12.66	9.40	7.47	6.47	5.14	3.75	3.26
Safflower	35.27	22.32	14.87	11.17	8.44	6.73	6.22	4.77	4.11	3.44
Sesame	41.44	24.83	16.80	11.91	8.91	7.19	6.25	4.95	4.16	3.43
Soybean	38.63	23.58	15.73	11.53	8.68	7.17	6.12	4.58	3.86	3.31
Sunflower	41.55	25.02	16.90	11.99	8.79	7.38	6.57	4.99	4.01	3.52
Walnut	33.72	21.20	14.59	10.51	8.21	6.71	5.76	4.80	3.99	3.36

Figure 8.11 Quenchalizer or quenchometer (left) and sample holder and probe (right)

can be a safer substitute for metal oil-based quench oils. Before a biobased oil could be considered a quench oil for a specific heat treat application, its cooling rate would need to be determined. An instrument called quenchalizer or quenchometer (Figure 8.11) maybe used to determine how quickly a probe heated to a standard temperature (typically 850 °C (1562 °F), cools with the test oil while at the same time monitoring the rise in the temperature of test oil.

Table 8.3 Key cooling characteristics of partially hydrogenated and winterized soybean oil

Key cooling characteristics				
Media	Partially hydrogenated winterized soybean oil			
Condition	40 °C Static			
File name	Soy oil cooling curve			
Probe	109			
Customer	UNI-NABL			
	Actual	Max.	Min.	Units of change
Maximum cooling rate	77.40	53.00	47.00	°C/s
Temp. at max. cooling rate	703.90	530.00	490.00	°C
Cooling rate at 300 °C	6.00	8.00	6.00	s
Time for immersion at 600 °C	6.00	14.00	12.00	s
Time for immersion at 400 °C	13.25	21.00	19.00	s
Time for immersion at 200 °C	49.00	55.00	50.00	s

In this test, a cooling time versus temperature pathway is established. This is directly proportional to physical properties such as the hardness obtainable upon quenching of a metal. The results obtained by this test may be used as a guide in heat treating oil selection or comparison of quench severities of different heat treating oils, new or used.

A quenchalizer sometimes called Quenchometer (Figure 8.11) uses a 2 l sample of oil and an instrumented probe which is heated to 850 °C, and submerged into the sample oil. The temperature of the oil and the probe are sensed and recorded to establish the cooling curve of the oil and the temperature change in the probe.

Table 8.3 shows an example of the cooling characteristics of soybean oil and the cooling curve for the same oil is shown in Figure 8.12.

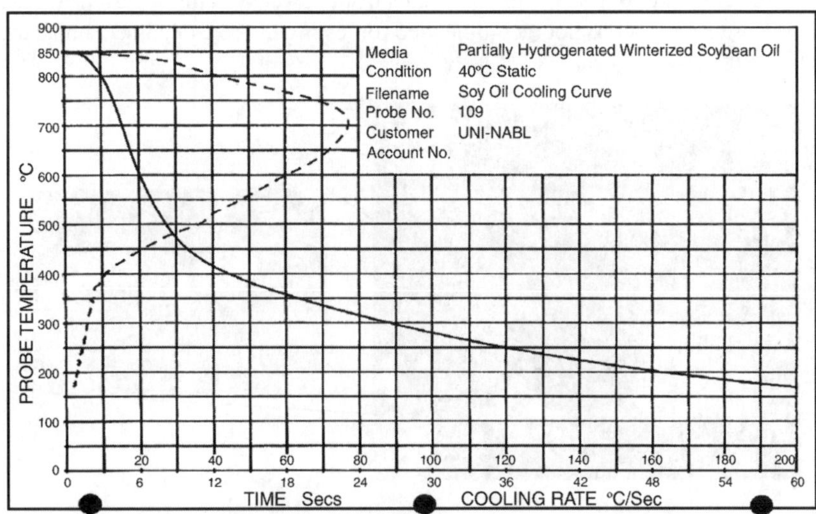

Figure 8.12 An example of a cooling curve for the soybean oil in Table 8.3

Figure 8.13 Pole-mounted soy oil-filled electric transformer (left), view of inside of the pole mounted transformer (center) and hydraulically operated boom truck (right)

8.4 Dielectric Properties

Hydraulic oils and transformer fluids are tested for their dielectric values. Hydraulic oils used in boom trucks for the maintenance of electric lines may come into contact with high voltage lines and high dielectric values, which indicate lower conductivity, are desirable (Figure 8.13).

The dielectric values of oils are tested according to ASTM D 2225, using a dielectric tester as shown in Figure 8.14. This test method covers a refereed and a routine procedure for determining the dielectric breakdown voltage of insulating liquids. These procedures are applicable to liquid petroleum oils, hydrocarbons, and oils commonly used as insulating and cooling media in cables, transformers, oil circuit breakers, and similar apparatus. The referee procedure, with modifications, is suitable for testing silicone fluids (see ASTM D 2225).

This procedure is also suitable for testing alkylbenzenes. The suitability of either the referee or the routine procedure for testing liquids having viscosities exceeding 900 cSt (mm^2/s) (5000 SUS) at 40 °C has not been determined. This test method is recommended for acceptance tests on unprocessed insulating liquids received from vendors in tank cars, tank trucks, and drums. It may also be used for the routine testing of liquids from power systems apparatus rated 230 kV and below. This test method is not recommended for testing filtered, degassed, and dehydrated

Figure 8.14 Dielectric fluid tester. See Plate 20 for the color figure

oils prior to and during the filling of power systems apparatus rated above 230 kV, or for testing samples of such oil from the apparatus after filling. Test Method D 1816 is more suitable for, and is recommended for, testing such oils.

The unit shown in Figure 8.14 is a Manual Liquid Dielectric Test Set Model Number L60 to run tests according to ASTM D 877, the test method for dielectric breakdown.

The test electrode is adjusted using the measuring rod, and the test cell is filled and placed into the instrument. The instrument is then set to rise at 3000 V/s. The voltage increases until it overcomes the dielectric, in this case the sample being tested. When failure occurs (dielectric overcome) the instrument will shut down and the breakdown voltage will be indicated on the meter. Soybean oil, for example, has shown a kV value of about 57 kV, which is higher than petroleum oils of the same viscosity in the 40–45 kV range.

8.5 Fluid Quality

There are many requirements to ensure fluid quality for use in varied applications. These quality requirements take a new twist when the base oil is not petroleum, like the biobased products. Since vegetable oil-based products can absorb water they require proper additives to improve hydrolytic stability, demulsibility, and long-term biostability. The ASTM D 6304 "Determination of Water Content in Petroleum Products, Lubricating Oils, and Additives by Coulometric Karl Fisher titration" can be used to determine the water content of the oil. Figure 8.15 shows the instrument used for Coulometric Karl Fisher.

Experience has shown that transportation of biobased base oils, especially neat vegetable oil during hot and humid summer months, could be problematic. The moisture in the atmosphere could be absorbed by the oil and increase the water content significantly. As a result, it is important to test the base oil for water content to ensure consistent quality.

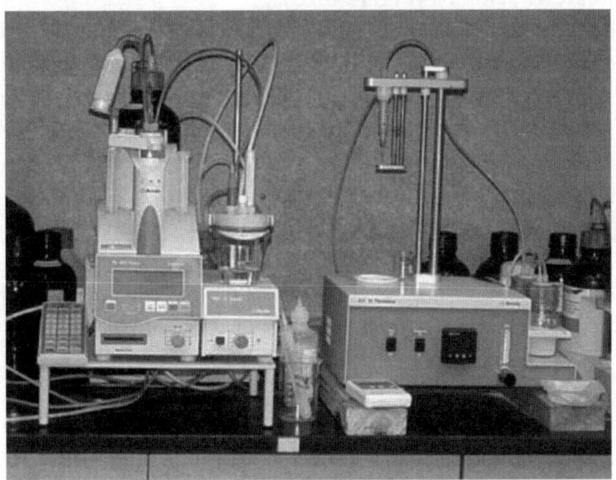

Figure 8.15 Instrument for coulometric Karl Fisher

8.6 Fluid Compatibility

Some original equipment manufacturers (OEMs) have established fluid compatibility requirements for vegetable oils in anticipation of their mixing into petroleum-based oils. For example, Caterpillar has defined a BF-1 Hydraulic Fluid Compatibility Method to ensure that biobased hydraulic oils could be mixed into petroleum-based hydraulic oils without a negative interaction. Similarly, Deere & Company has a test of compatibility called JDQ 23 Compatibility of Lubricating Oils.

Compatibility also refers to the compatibility of fluids with elastomeric seals and rubber hoses as well as the propensity to staining metals such as copper and brass in metalworking applications.

For elastomer compatibility there is the ASTM D 4289 method "Standard Test Method for Elastomer Compatibility of Lubricating Greases and Fluids" and the ASTM D 2240 "Rubber property-durometer hardness." In these test methods, the elastomeric material is exposed to the test fluid at elevated temperatures for an extended period of time. Then any change in volume due to swelling or shrinkage or change in the hardness the material is noted.

In this method, the test specimens of the elastomer materials are aged in the candidate oil for 100 hours up to 1000 hours depending on the specification at 100 °C. Some materials like 1E0724 urethane may be require to be aged at lower temperatures like 80 °C (176 °F). Initial properties of the elastomer samples including the hardness, tensile properties, elongation, and stress at 100% elongation must be evaluated and recorded before aging. Aging the test specimens in the candidate oil is then performed at the required temperature for the required period of time in hours by following the procedure outlined in ASTM D 4289. Then change in volume, loss in tensile strength, loss in elongation and residual elongation following ASTM D 4289, and the change in the Shore A hardness by following ASTM D 2240 are reported.

As an example, the allowed change in hardness, relative volume change, loss in tensile strength, loss in elongation, and residual elongation for the selected elastomeric materials are given in Table 8.4.

Manufacturers of laboratory balances have also designed attachments for measuring the weight of the elastomeric samples (Figure 8.16).

To determine the hardness or changes in the hardness of the material a durometer is used (Figure 8.17). The test is based on the ASTM D 2240 "Durometer Standard Test Method for Rubber Property – Durometer Hardness." This test method is based on the penetration of a specific type of indentor when forced into the material under specified conditions. The

Table 8.4 General specification requirements for elastomer compatibility

	Test temp. (°C)	Shore A hardness change (Pts)	Relative volume % change	Loss in tensile strength, % max	Loss in elongation, % max	Residual elongation, % min
HNBR(1 e2719)	100	+10/−15	−3/+20	50	50	80
NBR (1e0741)	100	"	"	"	"	"
FKM (1e0804)	100	"	"	"	"	"
AU (1e0724)	100	"	"	"	"	"
NEO-PRENE (1e0809)	100	"	"	"	"	"

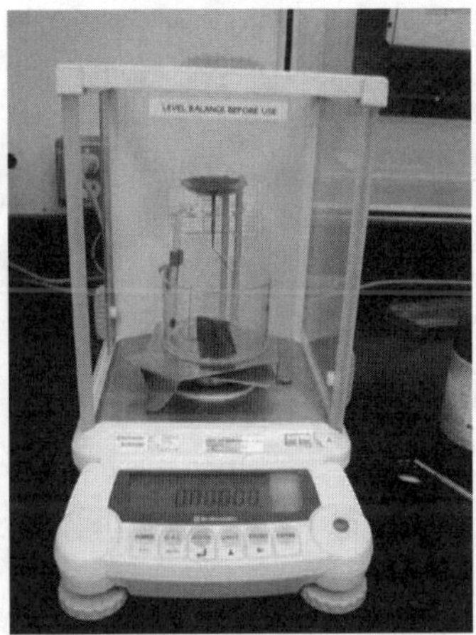

Figure 8.16 Measuring weight change due to elastomer swelling. See Plate 21 for the color figure

indentation hardness is inversely related to the penetration and is dependent on the elastic modulus and viscoelastic behavior of the material. **The geometry of the indentor and the applied force influence the measurements such that no simple relationship exists between the measurements obtained with one type of durometer and those obtained with another type of durometer or other instruments used for measuring hardness**. This test method is an empirical test intended primarily for control purposes. No simple relationship exists between indentation hardness determined by this test method and any fundamental property of the material tested.

The swelling and weight gains for the elastomers are performed by using special set ups for a balance to handle the very slight change in the mass of the sample being tested (Figure 8.18).

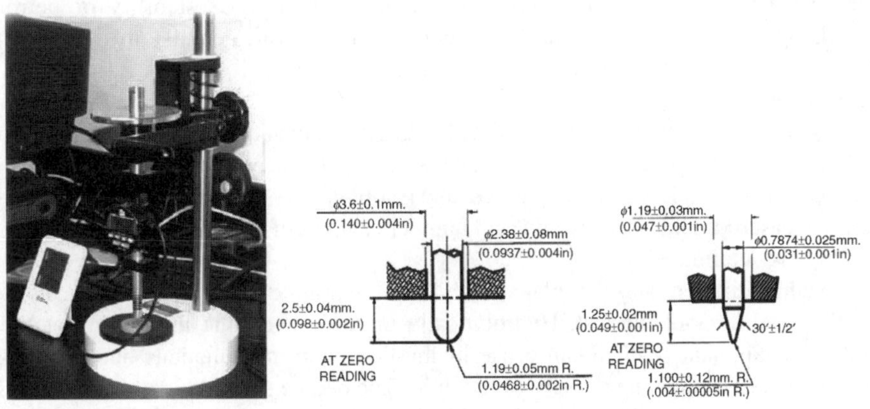

Figure 8.17 Durometer for testing hardness of elastomers ASTM D 2240 (left) and indentor types used for different materials

Figure 8.18 Special attachment for weighing rubber sample. See Plate 22 for the color figure

8.7 Hydrostatic Stability

Vegetable oils are prone to a stability break down when water is introduced into the product during use. The ASTM D 2619 "Hydrolytic Stability of Hydraulic Fluids" is often used to determine the stability of the oil during use, where the presence of water from the atmosphere or the rest of the machine is possible.

The test method covers the determination of the hydrolytic stability of petroleum or synthetic-based hydraulic fluids. Water-based or water-emulsion fluids can be evaluated by this test method but are run "as is." Additional water is not added to the 100 g sample.

The test involves a sample of 75 g of fluid plus 25 g of water and a copper test specimen being sealed in a pressure-type beverage bottle. The bottle is rotated, end for end, for 48 hours in an oven at 93 °C (199 °F). Layers are separated, and insolubles are weighed. The weight change of copper is measured. The viscosity and acid number changes of the fluid, and the acidity of the water layer is determined.

This method differentiates the relative stability of hydraulic fluids in the presence of water under the conditions of the test. Hydrolytically unstable hydraulic fluids form acidic and insoluble contaminants, which can cause hydraulic system malfunctions due to corrosion, valve sticking, or change in viscosity of the fluid. The degree of correlation between this test and service performance has not been fully determined. Figure 8.19 shows a hydrolytic stability tester, sometimes referred to as the Coke Bottle Test due to the use of older, heavier glass Coca Cola bottles as the standard test sample holder.

Figure 8.19 Hydrolytic stability tester ASTM D 2619

8.8 Demulsibility

This refers to the oil's ability to prevent water from mixing in. Using ASTM D 1401, the test sample shows separation of sufficient water having a volume of at least 37 ml observed before 20 minutes has elapsed (Figure 8.20). Emulsibility and demuslibility

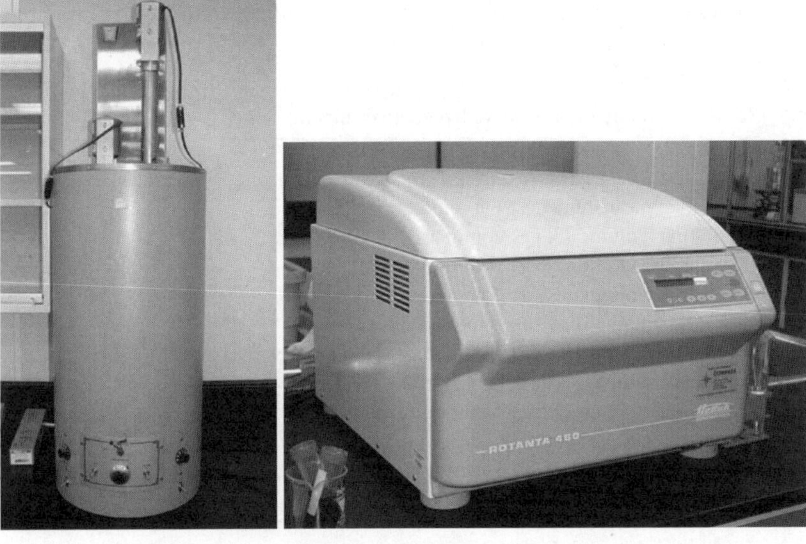

Figure 8.20 Demulsibility apparatus (left) and centrifuge

are two contradictory properties and in some applications the two need to be carefully balanced. For example, an engine oil may require enough emulsibility to absorb the moisture settled from the atmosphere which will have evaporated after the engine temperature reaches the boiling point of water. At the same time it needs a sufficient degree of demulsibility that a significant water leak from the engine cooling system would remain separate until discovered.

8.9 Oxidation Stability

For vegetable oil-based lubricants, it is most appropriate to use oxidation stability tests that are developed for vegetable oils. The common misconception that vegetable oils should also equal performance in tests that are developed for petroleum products could result in unduly eliminating some high performance vegetable oils from evaluation. For example, thin film oxygen uptake (TFOUT) is commonly used for evaluating base petroleum oils for engine oil formulation. TFOUT cannot differentiate between vegetable oils with different oxidation stability. Unpublished reports at the University of Northern Iowa's National Ag-Based Lubricant (UNI-NABL) Center have shown that vegetable oils of varied stability ranging in the oxidation stability instrument (OSI) from 7 to 500 hours, when tested in TFOUT, showed low values of 17 and 18 minutes. Similarly, petroleum oils will not perform the same way in the OSI as vegetable oils do, thus rendering the OSI useless for evaluating the oxidation stability of petroleum oils. As recommended earlier, the test of viscosity over 1000-hours in the ASTM D 7043 (formerly ASTM D 2271), along with OSI or AOM values, could provide a better assessment of stability of the oil for use in industrial lubricants. Table 8.5 presents OSI values for neat and the mixed versions of soybean oil and petroleum oil.

8.10 Oxidation Stability for Mineral Oils

Mineral oils oxidize when exposed to air and heat. As a rule of thumb, the rate of reaction of oil with oxygen increases by a factor of 2 for every 10 °C. In a handbook published by

Table 8.5 Oxidation of mixed vegetable-mineral oils

Sample	OSI (hours)
100 % Sample volume (5 g) commodity soybean oil	4.13
2/3 Sample volume (3.3 g) commodity soybean oil	1.95
1/2 Sample volume (2.5 g) commodity soybean oil	2.3
1/3 Sample volume (1.7 g) commodity soybean oil	2.68
100 % Sample volume (5 g) mineral oil	0
2/3 Sample volume (3.3 g) commodity soybean oil; with 1/3 sample volume (1.7 g) mineral oil	4.93
1/2 Sample volume (2.5 g) commodity soybean oil; with 1/2 sample volume (2.5 g) mineral oil	8.16
1/3 Sample volume (1.7 g) commodity soybean oil; with 2/3 sample volume (3.3 g) mineral oil	21.13

Figure 8.21 Rotary Bomb Oxidation Test (RBOT) fresh and spent copper catalysts. See Plate 23 for the color figure

Nynas Corporation [1] the oxidation mechanism is described in three stages: (1) creation of free radicals by heat, ultraviolet light, and/or mechanical shear; (2) creation of peroxides by the reaction of the free radicals with oxygen; and (3) the peroxide may further react and produce new radicals, alcohols, ketones, aldehydes, and acids (p.17). The Rotary Bomb Oxidation Test (RBOT) is a test commonly used for determining oxidation stability of petroleum products, and is finding popularity for testing the oxidation stability of biobased lubricants as well (Figure 8.21).

Based on the ASTM D 2893 test method "Oxidation Characteristics of Extreme Pressure Oils," the test method covers the determination of the oxidation characteristics of extreme-pressure fluid lubricants, gear oils, or mineral oils. The changes in the lubricant resulting from these test methods are not always necessarily associated with oxidation of the lubricant. Some changes may be due to thermal degradation. It utilizes a copper catalyst with the sample oil pressurized to a standard pressure and sealed in a bomb at a specified temperature. The drop in the pressure is used to determine the oxidation rate of the product.

8.10.1 Aromatic Content of Mineral Oils

The aromatic content of an oil can be determined using infrared radiation instrumentation methods to determine the percent of aromatic carbons, or using ASTM D 2140. These two methods do not return the same values.

8.11 Elemental Analysis

In biobased lubricants, a list of the elements included in the product is often obtained as per the ASTM D 5185 Elemental Analysis by the Induction Plasma Coupling (ICP) Method. Elements

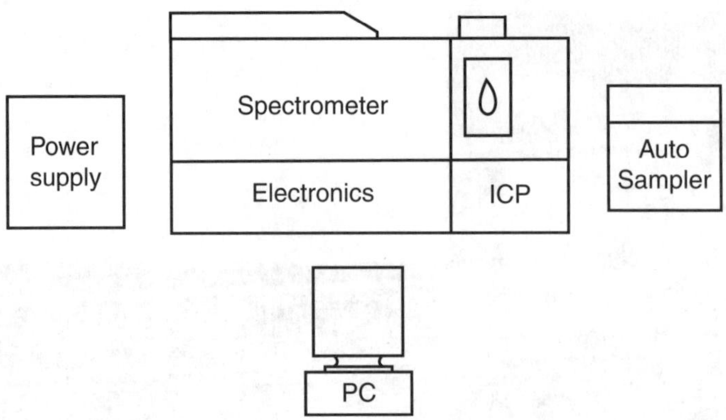

Figure 8.22 Simplified illustration of ICP

identified include: Ag, Al, Ba, Ca, Cd, Cr, Fe, Mg, Mn, Mo, Na, Ni, P, Pb, Sn, Ti, V, and Zn. Some specifications may not allow the use of certain metals, like sulfur and zinc which are considered undesirable, while they are commonly used in other applications for antiwear and as extreme pressure property enhancers. To determine the elemental content of an oil, commonly an ICP is used (Figure 8.22).

8.12 Cleanliness

Typically the finished lubricants products, especially hydraulic fluids, are tested using particle counters to determine the amount of particles in sizes of 2, 5, and 15 μm. Using the International Standard Organization's (ISO) method, an ISO Cleanliness level using three numbers is specified. These numbers are indicative of the fluid's contamination level based on the size of the particles in microns. The tests are performed based on the ASTM D 1401 method.

Table 8.6 shows the ISO 4460 code chart with number of particles per milliliter of the oil sample [4]. Currently, three two-digit numbers are used to express the ISO cleanliness requirement of an oil. The cleanliness requirements are dependent on the sensitivity of the components to contaminants and also to operating temperatures. Applications that involve very fine nozzles and high pressures require cleaner oils and may have ISO cleanliness requirements like 16/14/10.

Figure 8.23 shows a suggested guide for determining the cleanliness requirement of hydraulic oils based on the operational conditions.

While any particle counter may be used to determine the cleanliness of fluids, some instruments like one manufactured by Vickers, Inc. (currently Eaton in Eaton Corporation) have designed particle counters specifically for lubricants and hydraulic fluids (Figure 8.24). Using clean sample bottles, the oil is scanned for particles and a three-digit cleanliness code is automatically printed, making the machine useful and portable for in-plant preventive maintenance or for quality control.

Table 8.6 ISO 4406 Code chart particles per milliliter

Range code	More than	Up to/including
24	80 000	160 000
23	40 000	80 000
22	20 000	40 000
21	10 000	20 000
20	5000	10 000
19	2500	5000
18	1300	2500
17	640	1300
16	320	640
15	160	320
14	80	160
13	40	80
12	20	40
11	10	20
10	5	10
9	2.5	5
8	1.3	2.5
7	0.64	1.3
6	0.32	0.64

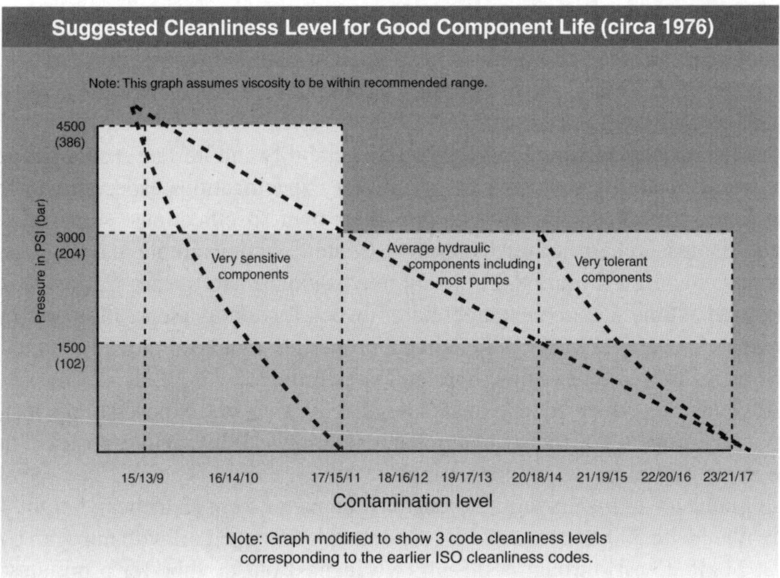

Figure 8.23 Suggested cleanliness levels for good components life

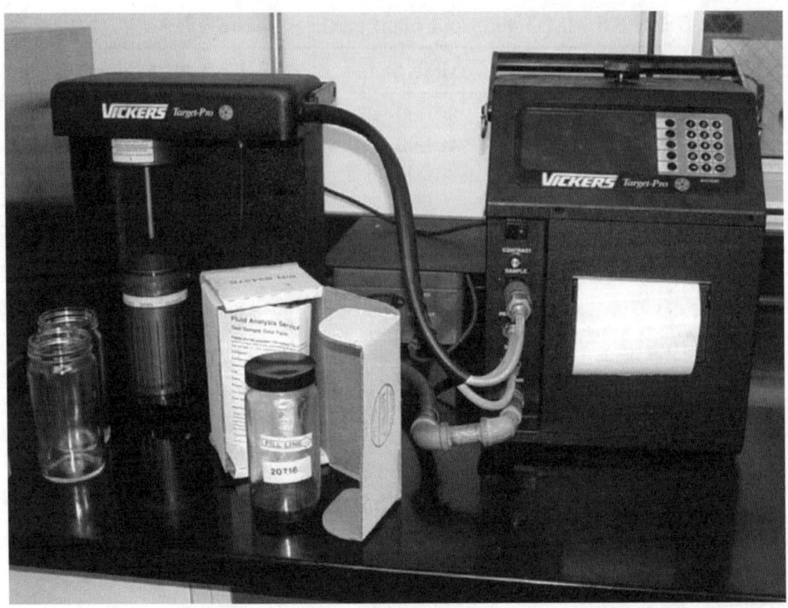

Figure 8.24 Portable particle counter designed to report ISO cleanliness numbers

8.13 Storage and Shipping Temperatures

Biobased lubricant products are generally shipped at ambient temperature. When ambient temperatures are within 10 °C above a product's pour point, provision must be made for heated shipping and storage.

There are existing test requirements for shelf life stability and storage at cold temperature or long-term storage stability for many of the military specifications especially for hydraulic fluids and fuels. Biobased lubricants require their own specifications, some of which are proposed by the researchers at the UNI-NABL Center. As an example, using programmable environmental chambers (Figure 8.25) the oil maybe exposed to freeze–thaw cycles or for an extended period of time at extreme temperatures to look for additive separation, sedimentation, color change, or changes in the physiochemical properties. One recommendation by the UNI-NABL Center is used as an example here and is as follows.

This test would be performed on two samples side by side, one exposed to open air and one topped off with nitrogen. The total acid number, oxidation stability, and viscosity of the sample should be tested and recorded before testing.

Two tall graduated cylinders with dimensions: diameter 5 cm (2 inches), height 45 cm (18 inches) are filled with 500 ml of sample oil. One sample is topped off with nitrogen by blowing nitrogen at 21 kPa (3 psi) through a 0.5 cm (1/6 inches) plastic tube for 5 minutes and then covered with high temperature oven-safe plastic. The samples are placed inside the environmental chamber and programmed to alternate 12 hours at each of test temperatures of 60 °C (140 °F) and 0 °C (32 °F) for a period of 10 days (240 hours).

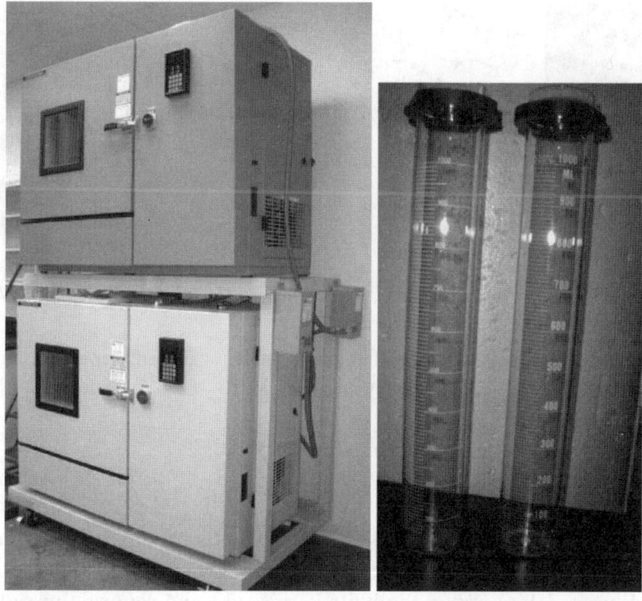

Figure 8.25 Typical environmental chambers for exposing sample oils to changing climates (left) and graduated glass cylinders

The samples should be inspected visually for any sign of settling or separation. Separation of any kind or settling of any component would be considered a failed product. Total acid number, oxidation stability, and viscosity should be tested following the testing. The following table is used for evaluation of these properties. The reporting would be based on a table as follows; and numbers are examples of recommended requirements for some biobased hydraulic oils.

8.14 Tribological Performance of Biobased Lubricants

When dealing with friction and wear, typically the goal is to determine the wear reduction or the metal to metal separation capacities of the lubricating oil. The most common types of metal to metal contact in the machinery take place between round on round, round on flat, or flat on flat surfaces. There are many instruments that are designed to measure wear or extreme pressure properties of lubricating oils for one or more of these three types of surface contacts. In addition to these instruments, there are a number of performance test rigs requiring the use of pumps, gear boxes, differential, and so on. Hydraulic pump tests, for example may be required for testing wear protection requirements of hydraulic oils. Those tests are typically more elaborate and take longer periods of time and are costly. The tests described below, however, are for screening purposes or for quality control. These include the following instruments.

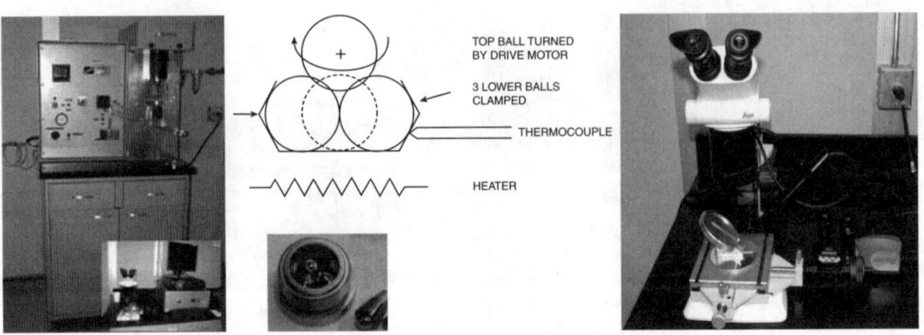

Figure 8.26 Four Ball Wear Tester (left) with microscope for scar diameter measurement. See Plate 24 for the color figure

8.14.1 Four Ball Wear Test: ASTM D 4172

In this test, three standard steel balls are locked in holding cup and a fourth ball placed in rotating chuck is allowed to rotate on three stationary balls for a period of 1 hour. The test oil temperature is kept at 93 °C (200 °F), and a weight of 40 kg is applied to the rotating ball. After the test, the diameter of the scar on the balls is measure and reported in millimeters. Some applications, for thinner fluids may specify a smaller load like 10 kg or 20 kg (Figure 8.26).

8.14.2 Four Ball Extreme Pressure Test

ASTM D 2596, ASTM D 2783, and ASTM D 2672. This test method is used to evaluate the extreme pressure properties of biobased lubricating oils. The required test conditions are as follows: variable kilogram loads, 20 °C (68 °F), 1750 rpm, and 10 second test durations (Figure 8.27).

At various kilogram loads, the test is run for 10 seconds, until the balls weld together due to the heat of friction. This "weld point" along with a load wear index are reported. In some biobased grease products, weld points of 500 kg are common.

8.14.3 Timken O.K. Load Test – ASTM D 2509

This test method can be used to evaluate the extreme pressure properties of biobased lubricating oils. The test specimens include a bearing race and block (Figure 8.28). Various loads are applied to the block while being pressed against the rotating bearing race submerged in the test oil. The scar on the block is rated as pass or fail based on the size of the scar and discoloration, using rating references.

8.14.4 FZG Rating

The Forschungstelle fur Zahnrader und Getriebebau (FZG) Visual Method, is intended to measure the scuffing load capacity of oils used to lubricate hardened steel gears. Scoring is a form of abrasive wear and is also included as a failure criterion in this test method. It is

Figure 8.27 Four ball extreme pressure tester – cup and chuck and a welded 4-ball sample. Source: ASTM International (2009), 100 Barr Harbor Drive, W. Conshohocken, PA 19428-2959, USA

Figure 8.28 Timken OK tester for extreme pressure (left) and new (top) used bottom tested specimens

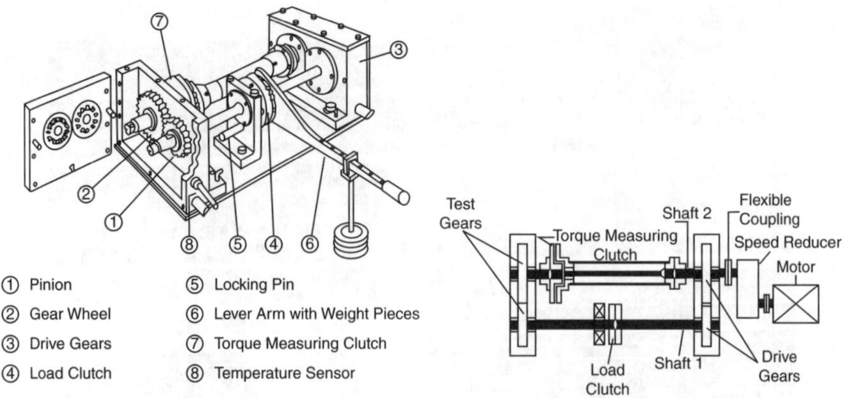

Figure 8.29 FZG gear oil test machine. Reprinted, with permission, from D5182-97 Standard Test Method for Evaluating the Scuffing Load Capacity of Oils (FZG Visual Method), Copyright ASTM International, 100 Barr Harbor Drive, West Conshohocken, PA 19428. A copy of the complete standard may be obtained from ASTM International, www.astm.org

primarily used to assess the resistance to scuffing of mild additive treated oils such as industrial gear oils, transmission fluids, and hydraulic fluids. High EP type oils are those oils meeting the requirements of API Gl-4 and GL-5 and generally exceed the capacity of the test rig. Therefore they cannot be differentiated with this test method.

When properly formulated, vegetable oils have performed well in the FZG tests. This test method is used to screen the scuffing load capacity of oils used to lubricate spur and helical (parallel axis) gear units. The test is considered an important criterion in the specification of many gear oils and transmission fluids.

The FZG is operated at a constant speed (1450 rpm) for a fixed period (21 700 revolutions for approximately 15 minutes) at successively increasing loads until the failure criterion is reached. The initial oil temperature is 90 °C (194 °F) beginning at load stage four (Figure 8.29). The test gears are examined initially and after the prescribed duration at each load stage for cumulative damage (scuffing) to the gear tooth flanks.

In addition to these test methods there are numerous tests that deal with friction and wear performance of oils in specific applications. For example, the Pin and Vee Block test or Tapping Torques test (both explained later) are used for testing the performance of metalworking fluids. Hydraulic pump test stands and gear and differential test units, or engines test rigs, are required for testing the performance of hydraulic oils, gear oils or engine oils respectively.

8.15 Metalworking Fluids

Metalworking fluids encompass a whole area of science and art of fluid preparation beyond the scope of this book. There are, however, biobased metalworking fluids on the market and some comparative data are presented here for reference purposes.

There are many different types of metalworking fluids based on their use, and their requirements vary considerably. Some could be very simple and even pure water or pure vegetable oil could be used for cooling or lubrication, respectively. Others, however, are complicated and require extensive and complicated formulations. Generally, the metalworking fluids are divided into two distinct categories: (1) straight oils (sometime referred to as NEAT oil) which come in oils of various viscosities; and (2) coolants (typically water-soluble oils) which come in oil concentrate but are mixed with water at small percentages of 5–10% oil to 95–90% water. Each of these types of fluids requires a different formulation and a different level of maintenance during use.

Straight oils are somewhat easier to maintain if formulated properly. A Way Oil, for example, is applied to the way of a moving head of metal cutting machine. As long as it does not thicken up, does not stain the materials being worked on, and does not interact with other oils or coolants, a Way Oil would work rather trouble free. The key for a successful product is in its formulation, to include the right amount of antiwear, anti-rust, and other property enhancer chemicals.

Coolants or water-emulsified metalworking fluids, on the other hand, are more complicated in both their formulation and manufacturing as well as in their use. Coolants are typically formulated in a concentrated form but mixed with water at the point of usage. The oil requires the right amount of emulsifier to provide a stable emulsion. Failure to formulate properly will result in oil-water separation. Additionally, due to the high level of water in the coolant along with the heat from the machine, the coolant provides a suitable environment for bacteria to grow. So, metalworking coolants require bactericides, fungicides, and other chemicals to prevent bacterial growth. The presence of bactericides in the coolant is not healthy for the workers. Cases of dermatitis (skin conditions) and cases of lung ailments (when the mist is inhaled) can be prevalent.

Furthermore, the operator would have to be educated about the maintenance of the coolant. Trained operators monitor the pH of the coolant consistently to ensure it remains in the 8–10 range, which does not promote bacterial growth. When formulated properly, the coolant has a white, milky appearance and will remain stable for a long period. Figure 8.30 shows test tubes containing straight oils and water emulsified coolants.

Other than proper formulation at the source and proper maintenance in use, there are numerous other factors that could cause coolants to fail. The hardness or softness of the water being used, for example, can impact the effectiveness of the emulsifier at the point of usage. In manufacturing, too, the order of mixing the additives and the temperature of each component when mixed could impact the stability of the product. Furthermore, during use, the water evaporates and needs replenishment, but the addition of water dilutes the additives of the coolant concentrate resulting in the need for tankside additives to be added.

Metalworking fluids, nevertheless, present a great opportunity for vegetable oils. An ideal metalworking fluid, among many other attributes, would have low volatility, high thin film strength, high flash and fire points, high viscosity index, and high adherence to metal surfaces due to their polarity. All of these features are the natural properties of vegetable oils. As a result, vegetable oils tested against conventional metalworking fluids perform superbly better than conventional metalworking fluids. The challenge, of course, is to ensure that the straight oil is oxidatively stable enough to survive the environment of the metalworking fluids, and the coolant is formulated to ensure that bacteria cannot survive in it. These two parameters are simple to understand yet difficult to achieve because of the

Figure 8.30 Test tubes containing soybean oil-based straight metalworking fluids and water emulsified coolants. See Plate 25 for the color figure

variability of machine environments and the attention of the operator to maintenance. Figures 8.31 and 8.32 show the use of soybean oil-based straight oil in metal spinning and coolant in cast iron metalworking applications.

One of the methods for testing the effectiveness of metalworking fluids is the use of Pin & Vee Block machine (ASTM D 3233). In this test, two V slotted specimens are pressed against a round stock. While the round Pin is rotated, more force is applied to the V blocks and the torque required to turn the round stock is monitored. The test specimens are placed in a container that holds the test fluid and is heated to 75 °C (167 °F). A toothed wheel rotates while each wheel tooth causes the V blocks to be pressed further against the round Pin. Figure 8.33 shows the Pin & Vee tester with pin and V block specimens.

This test is very severe and the friction between the flat V surfaces and the rotating Pin increases as the forces applied to the V blocks increase. At higher applied forces, the friction and heat increases enough to momentarily weld the V block surfaces to the pin and then break instantaneously resulting in gouging of the metal surfaces. Figures 8.34 and 8.35 show comparisons of metalworking straight oil and coolant made from vegetable oils compared with conventional mineral-based fluids.

The Pin & Vee tests of commodity crude soybean oil tested against refined high oleic soybean oil are shown in Figure 8.30. These oils were not treated with any additive to show base

Figure 8.31 Soybean oil-based coolant in cutting application with cast iron engine blocks

Figure 8.32 Soybean oil-based straight oil in metal spinning (left) and screw cutting applications

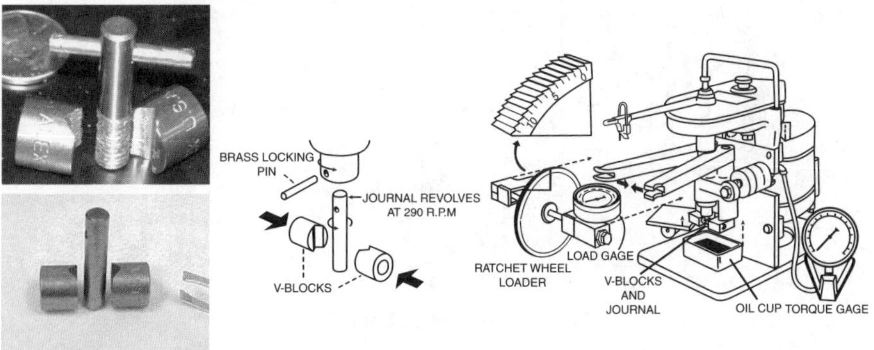

Figure 8.33 Schematic showing how pin and V block machine works (right). Courtesy of the Falex Corporation, Illinois, USA; and specimens used (top left) and unused

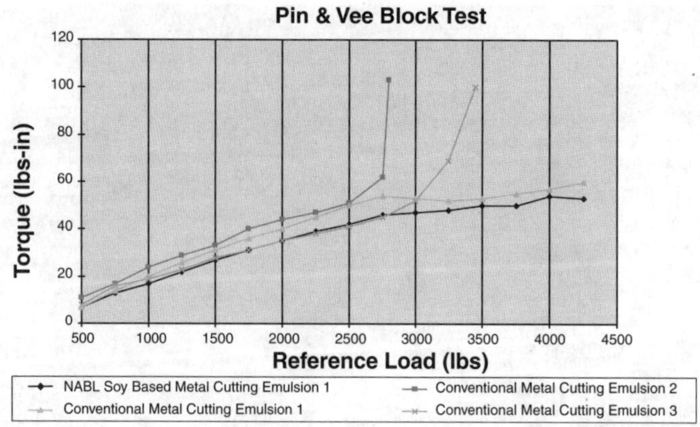

Figure 8.34 Comparison of a conventional straight oil with a soy oil-based straight oil

line information. Due to the high lubricity of vegetable oils these higher force and low torque readings are expected. The challenge for the formulator is to ensure that the oil is stable enough for repeated and extended use.

Figure 8.36 shows a Tapping Torque tester. Using standard tap and test specimen the torque is measured while the threads are being tapped. Oils with higher lubricity would result in lower torque for the tap. Biobased oils in general perform very well in this test.

Biobased metalworking fluids offer many advantages and are bound to play an important role in the metalworking industry. These products are more complicated to formulate and require considerable resources to ensure the proper maintenance during usage. With proper formulation by the manufacturer and proper maintenance by the user, biobased oils and coolants can increase the efficiency of cutting tools and the competitiveness of the metalworking industry.

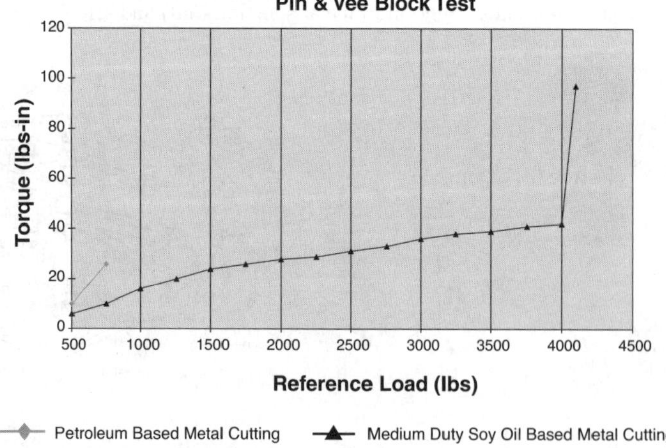

Figure 8.35 Comparison of three conventional coolants with a soy oil-based coolant

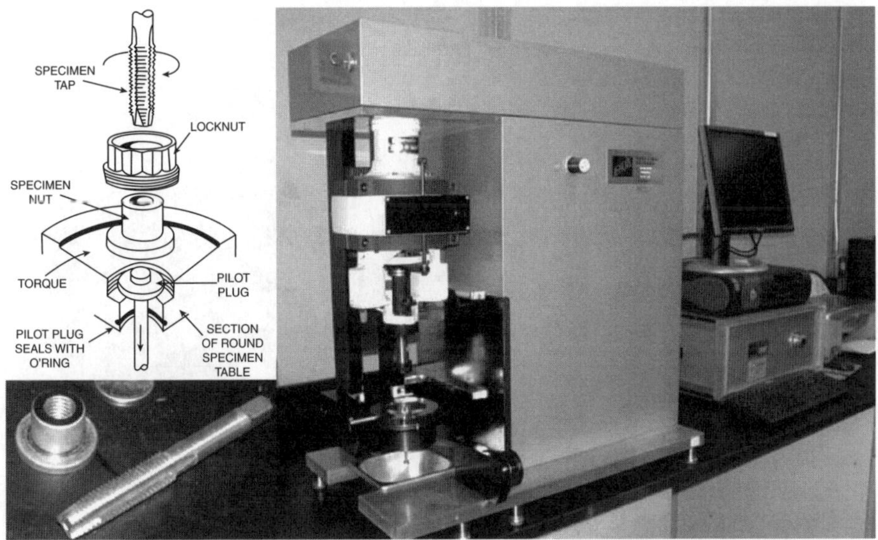

Figure 8.36 Tapping Torque machine (right) and standard specimen and schematic of the specimen (left). Reprinted, with permission, from D5619-00 Standard Test Method for Comparing Metal Removal Fluids Using the Tapping Torque Test Machine, Copyright ASTM International, 100 Barr Harbor Drive, West Conshohocken, PA 19428. A copy of the complete standard may be obtained from ASTM International, www.astm.org

8.16 Biobased Engine Oils [5]

Creating biobased engine oil has been like the search for the Holy Grail for researchers and product developers. For the past 20 years various small and large entities have been experimenting with biobased engine oils. Large entities abandoned the search after numerous trials and being faced with the ever-changing specifications. Other smaller groups created ventures and drove a truck filled with biobased engine oil, licking the dipstick at various stops to make a point for "environmental friendliness" of the product. Yet others in Europe experimented with the Continuous Oil Recycling concept in diesel passenger cars. While complex esters derived from vegetable oils can become economical and stable enough to handle the engine environment, many of the current vegetable oil-based technologies have lacked the necessary stability and cold temperature flowability to perform in the engines the way mature and perfected petroleum engine oils do.

As of the time of writing, claims for biobased engine oils are viewed with suspicion. Some have included small quantities of biobased oils in conventional engine oils and claimed the oil to be biobased. It is likely that eventually the biobased engine oils will be based on biobased-derived synthetic base oils. Using vegetable oils, even in small quantities could still result in oxidation of the oil and possible formation of polymer films within the oil galleries or exposed surfaces of the oil pan.

An approach by the UNI-NABL Center that involves a new approach to using biobased engine oils was described in the May 2009 issue of the *OEM Off-Highway* magazine (Figure 8.37) [5].

Continuous Oil Recycling System (CORS) was born from an after-market product that was added to a diesel engine to continuously steal a small amount of engine oil and feed it into the fuel

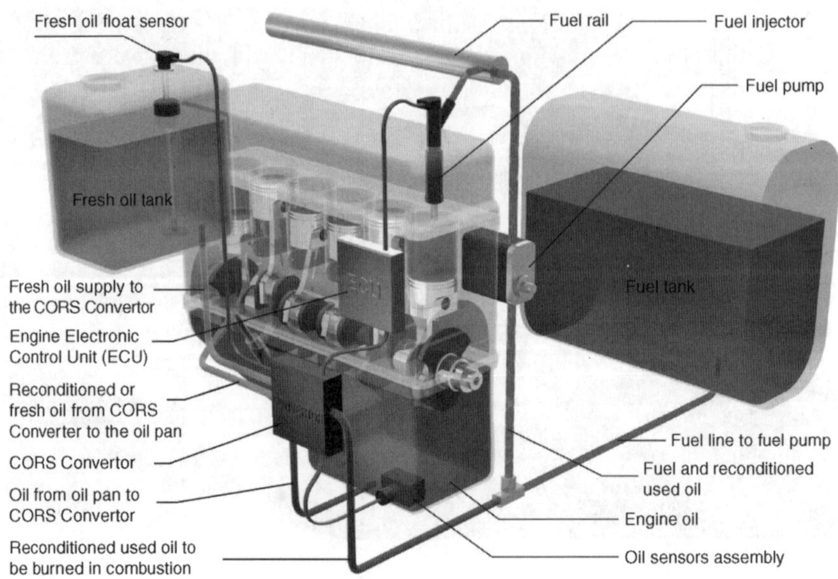

Figure 8.37 Conceptual representation of the CORS components plus the engine ECU and injector. See Plate 26 for the color figure

line of a diesel engine. The driver would then add on a small quantity of oil to the engine at every refueling. An add-on reservoir containing engine oil could also be used to refurbish the crank case with fresh engine oil. As a result, after so many refuelings, the operator would purchase oil for the add-on oil tank. The main problem with this approach was that the burning of used engine oil resulted in a significant, negative impact on the already hard to achieve emission standards.

UNI-NABL researchers modified the CORS concept to use vegetable oil-based engine oils and created a sophisticated "black box" called the **CORS Convertor**. The CORS Convertor not only communicates with the main processor of the engine, but it also has its own sensors and transducers to monitor the conditions of the engine oil and engine fuel before introducing the oil into the fuel line. The CORS Convertor is an add-on unit that conceptually resides between the crank case and the fuel injectors and ensures that the recycling of the oil as fuel is performed in a way that is beneficial to the engine.

CORS Convertor utilizes an array of sensors plugged into the crank case to monitor the physiochemical conditions of the engine oil, which is made of a mixture of modified vegetable oils and a minimal amount of performance-enhancing additives. The CORS Convertor continuously draws oil samples from the crank case and in addition to evaluating the oil, it conditions it as needed, to be used as fuel. In effect the CORS Convertor is expected to act like a human kidney that ensures what is fed into the fuel system is cleansed of harmful constituents and is only sent to the fuel system when the engine conditions indicate the best time to use this oil as fuel. For example, a cold engine would not receive vegetable-based engine oil into its fuel system, and the amount of oil that is fed into the fuel system is varied based on the ability of the engine's combustion process to provide the cleanest combustion possible.

Whereas the original idea of feeding used petroleum-based engine oil into the fuel system had a negative impact on the emissions, the CORS combined with the Convertor can actually improve exhaust emissions when using biobased oils.

8.16.1 Stationary Diesel Engines for CORS

The initial focus for the CORS is stationary diesel engines where the addition of an oil tank next to the fuel tank does not present the logistic problems of dealing with mobile equipment. To test the concept, however, three John Deere diesel engines were prepared and instrumented using an Eddy Current dynamometer to test the engine under various load conditions. Figures 8.38 and 8.39 show the engine, dynamometer, and monitoring equipment associated with the data collection system.

The engine and dynamometer sensor and transducers were interfaced with a data monitoring system to allow monitoring of the engine performance. For testing the engine, the maximum load and horse power setting was selected in order to accelerate the degradation of the oil. The four cylinder engine was rated at 120 Hp and the test parameters were set to simulate the full load at about 120 Hp with a range of ±5 Hp.

It is assumed that data gathered using these diesel engines will be transferrable to the stationary engines used as back up or small scale electric generators for utility companies, hospitals, and any place using diesel engine for electric generation or for pumping.

When the engine starts, regular diesel engine fuel is used without any mixing with the engine oil. The CORS Convertor in the meantime, draws small quantities of the oil from the engine via a positive displacement pump. The physiochemical properties of this oil are constantly monitored by an array of sensors placed in the crank case oil. The CORS Convertor receives

Figure 8.38 120 HP John Deere diesel engine

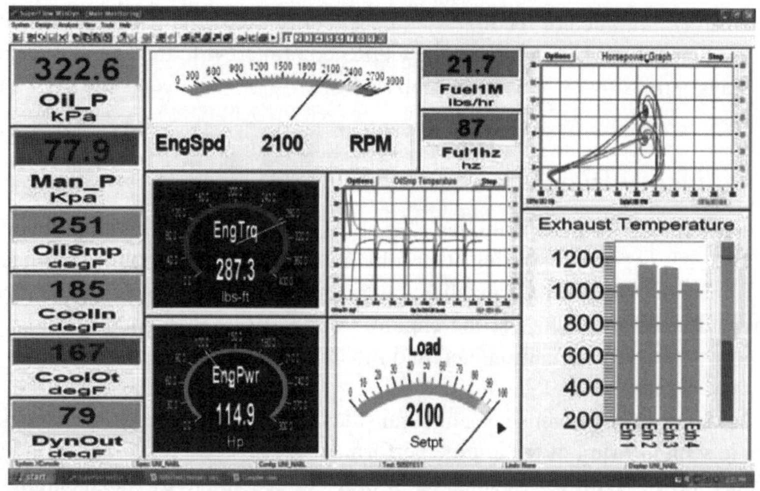

Figure 8.39 A screen shot of the parameters set for the engine tests. See Plate 27 for the color figure

the oil and depending on its conditions sends it through a number of conditioning steps based on a design that could include physical and chemical filtration, addition of chemical catalysts, heating or cooling, and so on. Some of this oil will remain in the CORS Convertor in a small reservoir ready to be sent into the fuel system while some of it might return back to the engine crank case if not needed at that moment.

The CORS Convertor has three oil lines attached to it; one for bringing in the engine oil for analysis and conditioning, one from the fresh oil reservoir for preparation for injection into the crank case to replenish the used oil, and one to the crank case for injection of either fresh oil, excess re-conditioned oil, or a mixture of fresh and reconditioned oil back into the crank case. When the data from the engine electronic control unit (ECU) and the engine sensors from CORS indicate that the conditions are right to feed the oil into the fuel, a small quantity, currently not exceeding average 2% of the weight of the fuel being consumed, is sent into the inlet side of the fuel injector pump. This creates a B-2 biodiesel fuel and since the vegetable oils have shown to improve the lubricity of low sulfur fuels, the result is a tribologically more effective fuel. The conditioned oil is void of most of the impurities often found in the used engine oils. Also, the oil is conditioned to proper temperature and at 2% level, is highly diluted in the fuel. This results in minimal impact on the power output of the engine due to the naturally lower BTU/volume content of vegetable oils compared to No. 2 diesel fuel.

8.16.2 Test Results

A review of several diesel engines, commonly used for agricultural machinery, was used to determine the time residency of the oil in the crankcase at different oil consumption rates. Table 8.7 shows the reservoir sizes and the percentage of the oil that has to be removed from the crank case in order to meet the desired residency time. It shows the time

Table 8.7 Specification requirements for changes in OSI, TAN, and viscosity after storage test

Property	Initial reading	After 240-hour test reading	Difference	Pass With nitrogen barrier	Pass Without nitrogen barrier
OSI (hours)			Δ OSI	1%<	2%<
total acid number (mg KOH/g)			Δ TAN	1%<	2%<
Viscosity in cSt at 40 °C (104 °F)			Δ Viscosity	1%<	2%<

residency of the oil in the engine based on the percentage of the oil taken away to be burned into the fuel.

In order to test the concept, UNI-NABL researchers prepared a pure vegetable base oil with a viscosity of 107 cSt at 40 °C as a replacement for 10W-40 diesel engine oil that has a viscosity of 120 cSt at 40 °C. The reason for using a lower viscosity vegetable oil-based engine oil was because the vegetable base oil has a much higher VI of about 220; as compared to equivalent petroleum oil with a VI of about 100. This allows the use of thinner oil and less starting torque requirement, while maintaining the viscosity of the oil at the operating temperatures (Figure 8.40).

The viscosity of the oil was checked every hour, knowing that an increase in viscosity indicates oxidation. After some initial shearing of the oil, the viscosity was stable for about 10 hours and then began to increase after about 10 hours of operation. During this test, the CORS was turned off and no oil was being removed from the crank case for burning as fuel. The untreated vegetable oil showed stable viscosity for about 10 hours at full load before the viscosity began to increase rapidly and the test was terminated after 14 hours. This viscosity increase was not unexpected as similar viscosity changes had been observed in high pressure-high temperature hydraulic pump tests but after longer hours of operation. Since the engine was running at full load, it is anticipated that a lower level of engine load could increase this operational performance to several hours more than the observed 10 hours. The time to rapid increase in viscosity corresponds to the OSI time for the oil. For comparison, Table 8.8 shows the hypothetical burn rates for four diesel engines and their associated drain (oil change) rate in hours.

Table 8.8 Burn rate and oil replacement in hours for four agricultural diesel engines

Engine Data	Agriculture engine 1	Agriculture engine 2	Agriculture engine 3	Agriculture engine 4
Engine power (kW)	93	138	150	360
Oil capacity (liters)	13	20	28	42
Engine fuel use (kg/h)	20	31	34	67
Drain rate at 0.25% burn rate (hours)	225	232	296	226
Drain rate at 1% burn rate (hours)	113	116	148	113
Drain rate at 5% burn rate (hours)	11	12	15	11
Drain rate at 10% burn rate (hours)	6	6	7	6

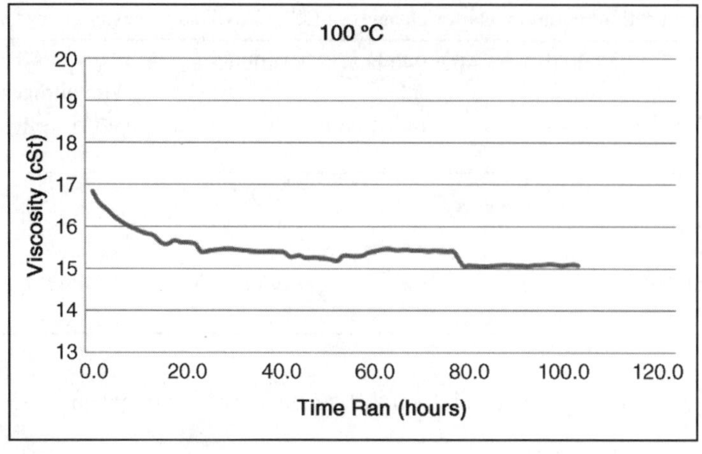

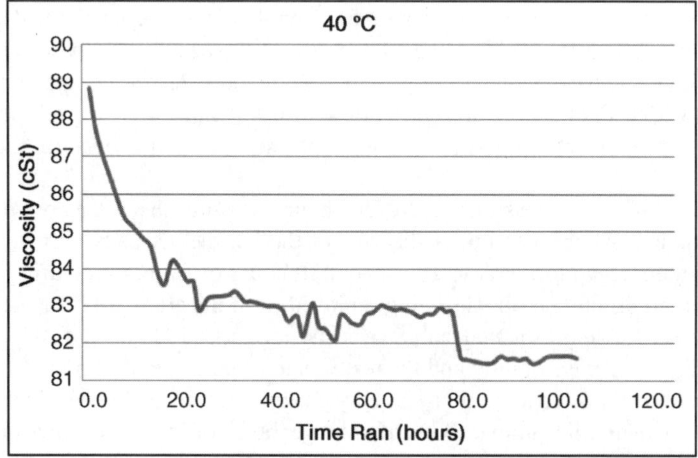

Figure 8.40 Viscosity values at 40 °C and at 100 °C for 100 hours at full load

Next, the same mixture of 75–25 vegetable oils/petroleum was tested with the engine at full load, and the viscosity of the oil remained flat for the entire 100 hours with CORS on. All other variables in the engine indicated that the mixture could be run in the engine indefinitely as long as an oil volume equal to 2% of the fuel consumption is removed and replaced with fresh oil. Since similar tests have been performed on the fuel using 2% purified vegetable oil in the fuel system without any impact on engine performance, the researchers were convinced that CORS is viable and will run a clean engine and a clean combustion indefinitely. These are still considered experimental products at the time of this publication. However, developmental activities are on-going for optimizing the size and components of the CORS Convertor, improving the accuracy of the sensor's assembly, and optimizing the algorithm used on the CORS processor. The engine oil is undergoing further improvements to replace the 25% petroleum portion of the blend with additives in the 75% vegetable oil portion. While the engine tests are continuing in the laboratory, field test sites are being run for long-term tests.

Economically, when the oil is consumed at the rate of 2% of the fuel consumption, its residency in the engine would range from 57 to 64 hours depending on the size of the engine and crank case, as shown earlier in Table 8.7. Since the residency of the oil is limited, there is reasonably little need, and as a result little cost, for the performance-enhancing additives. Because the oil would be used as lubricant first and then would be burned with almost par value for the fuel, the cost of the lubrication could almost be eliminated. The inconvenience of having to resupply the fresh oil tank would be compensated with the elimination for the need to change the engine oil. Understandably, CORS is still in the developmental stage, but it could deliver the first real engine lubricant technology based on vegetable oils.

References

1. Nynas, A.B. (2009) *Naphthenic Specialty Oils for Greases Handbook*, Nynas AB, Sweden, www.Nynas.com.
2. Facina, O.O. and Colley, Z. (2008) Viscosity and specific heat of vegetable oils as a function of temperature: 35 °C to 180 °C. *International Journal of Food Properties*, **11**, 738–746.
3. Elastomer Compatibility Requirements, Caterpillar Corporation BF-1 Specification for Biodegradable Hydraulic Fluids. Published by the Caterpillar Corporation (1996).
4. Vickers Incorporated (1992) "Vickers Industrial Hydraulic Manual"; Vickers Training Center, Rochester Hills Michigan, USA.
5. Honary, L.A. (2009) OEM biobased engine oils, *Off-Highway Magazine*, Sygnus, Fort Atkinson, WI USA.

9

Biobased and Petroleum-Based Greases

Grease is an effective means of delivering lubrication to the machine component being lubricated. Although liquid lubricants flow easily, they require a reservoir to contain their volume. Solid lubricants on the other hand require direct contact at the point of lubrication in order to effectively deliver lubrication. Greases are semi-solid and can deliver the benefits of liquid lubricants without requiring a reservoir, and also the benefits of solid lubricants by maintaining their body structure. In applications like the wheel bearing of an automobile where excessive heat is generated, liquid lubricants would thin down and can leak out of the bearing seals. The wheel bearing is a good example for using grease.

Grease as a semi-solid lubricant is made of a mixture of soap (semi-solid) and oil (liquid). Compared to liquid lubricants, grease maintains its body within the wheel bearing, filling all the bearing cavities while at the same time allowing the oil in the grease to lubricate the bearing.

By definition, grease is a solid or semi-solid fluid made of a thickening agent in a liquid lubricant and other ingredients that impart special properties (ASTM D 4175). The body of the grease is formed by the soap, which forms a matrix structure similar in detail to that of a sponge, thus allowing the voids within the matrix to be filled with the lubricating oil. The soap provides the body to keep the grease in place while the oil delivers the lubricity. An important benefit of the soap in the grease is that it provides the ability to seal and prevent the oil from leaking out or dirt from getting in. The soap's matrix acts as a reservoir for the oil in providing delivery of the lubricants and performance additives at the point of contact. As indicated earlier, grease is also made up of chemical performance enhancers (additives). The additives are used to match the performance requirements and could be used for wear protection, extreme pressure, corrosion resistance, antirust, and tackiness.

9.1 How to Make Soap

Simply put, grease is a mixture of soap and oil. Soaps are commonly familiar due to their prevalence in daily use as hygienic products. Before explaining how to make soap, it's useful to learn about some of the relevant terminology used in the grease industry.

Biobased Lubricants and Greases: Technology and Products, First Edition. Lou A.T. Honary and Erwin Richter

(1) **Soap.** The product of the reaction of a [fatty] ACID with an alkali (a strong BASE).
(2) **Salt.** ACIDS and BASES neutralize each other. A salt and water are the products of a chemical reaction between an acid and a base. When the acid is a *fatty acid*, then the product of the neutralization is called *soap*. Salts tend to be insoluble in oils, whereas soaps are reasonably soluble in oils.
(3) **Fatty acids.** Triglycerides of fatty acids make up the major components of vegetable oils, which are typically in *liquid* form. Each fatty acid is comprises a chain of carbon atoms of varying length. In addition to being bonded to itself, carbon atoms in the chain are bonded to hydrogen or other simple atoms (i.e. oxygen, nitrogen, sulfur).
(4) **Fats.** Fats are made up of carbon chains and hydrogen. In general fats are derived from animals and contain fatty acid chains that are saturated with hydrogen and thus they tend to be solids.
(5) **Saponification.** The process of forming soap by reacting (cooking) the fat (an acid) with a strong base to separate the glycerin from the fatty acids and forming soap with the resulting fatty acids.
(6) **Tallow.** The fat from animals; specifically, rendered fat from beef or mutton. If kept away from oxygen (air), tallow can be stored for a reasonably long time without refrigeration.
(7) **Tall oil.** Originating from the Swedish word *talloja*, or *pine oil,* it is a byproduct of wood pulp manufacturing (also called liquid rosin). Crude tall oil contains fatty acids like palmitic, oleic, and linolenic. Tall oil fatty acid (TOFA) is generally a cheaper alternative to tallow fatty acids [1].
(8) **Lye.** Also known as caustic soda, lye is *sodium hydroxide* (NaOH). It's readily available and has many uses. When mixed with water, a highly exothermic reaction occurs. Lye has a strong affinity for water and can draw moisture from the atmosphere.

Table 9.1 presents the fatty acid profile of tallow, butterfat, and lard [2].

In the past, making soap, or what is called *lye soap,* was a well-known process for both human usage and technology. Heating together animal fat (as acid) with the ashes from a wood stove (as base) made soap. These ashes contained a strong base (either sodium or potassium hydroxide). Initially, no attempt was made to separate the glycerol from the fat, thus resulting in referring to the soap as "glycerol soap." This type of soap did not have the consistency and performance of today's soap but was still suitable for laundry and bathing. In this case, the fat supplied the fatty acids and the ashes supplied the base. Using flower petals or scent essence provided the soap's fragrance. The presence of glycerin in the soap results in the soap having the affinity to moisturize the skin, as glycerin is highly hydrophilic. There is a general agreement that, on the skin, the soap residue absorbs moisture from the air making the skin feel softer, although there could be other reasons for this skin softness.

The presence of glycerol in the grease could impact the performance of the grease negatively, especially in cold temperatures. As the separation of glycerol from the fatty acid (hydrolysis) is an important factor in the final products, different chemical and enzymatic approaches are used. If the soybean is reacted with methanol to separate the glycerol from the fatty acid, then methyl ester is produced. Methyl ester is used as biodiesel but is of little value for grease making. Methyl esters have viscosities in the range of 2–4 cSt at 40 °C (104 °F), which is too thin for many lubricant uses. Furthermore, high solvency causes swelling of rubber hoses and seals. Other industrial process for hydrolyzing the fats and oils involve the Colgate–Emery process, or variations of it, which uses high pressure steam. Hammond *et al.*

Table 9.1 Fatty acid profiles of tallow, fat, and lard

Oil or fat	Unsat./sat. fat ratio	Saturated					Monounsaturated	Polyunsaturated	
		Capric acid	Lauric acid	Myristic acid	Palmitic acid	Stearic acid	Oleic acid	Linoleic acid ($\omega 6$)	Alpha linolenic acid ($\omega 3$)
		C10:0	C12:0	C14:0	C16:0	C18:0	C18:1	C18:2	C18:3
Beef tallow	0.9	—	—	3	24	19	43	3	1
Butterfat (cow)	0.5	3	3	11	27	12	29	2	1
Butterfat (goat)	0.5	7	3	9	25	12	27	3	1
Butterfat (human)	1	2	5	8	25	8	35	9	1
Lard (Pork fat)	1.2	—	—	2	26	14	44	10	—

(Patent #5089403) [3] describes a process for enzymatic hydrolysis of fatty acid triglycerides with lipase. Lipase is a known enzyme in the human digestive system, necessary for the absorption and digestion of nutrients in the intestines. This digestive enzyme is responsible for breaking down lipids (fats), in particular triglycerides, which are fatty substances in the body that come from fat in the diet.

The Hammond invention refers to the use of moistened, dehulled oat seed or oat caryopses. By mixing the caryopses with the oil, the fatty acids dissolve in the oil phase and the glycerol is absorbed in to the caryopses. This facilitates the separation of glycerol from the free fatty acid.

To make industrial soap as an oil-holding matrix (grease), it is more desirable to remove the glycerol from the fatty acid. The presence of the glycerol in the grease would increase its affinity for water and would negatively impact its flow ability at cold temperatures.

The most widely used and produced acids in the United States are inorganic acids or mineral acids. They do not occur in nature and must be manufactured. Fatty acids on the other hand are organic compounds produced by both plant and animals. An important [fatty] acid for making soap-based grease is stearic acid, which has a carbon chain of 18 carbons. This is a saturated fatty acid and is more suitable (due to its relative weakness as an acid) in the formulation of industrial greases and personal hygienic products. The University of Oregon Chemistry Laboratory resource center has a simple illustration of the stearic acid with OH (hydrophilic head highlighted away from the hydrophobic carbon chain tail), as shown in Figure 9.1.

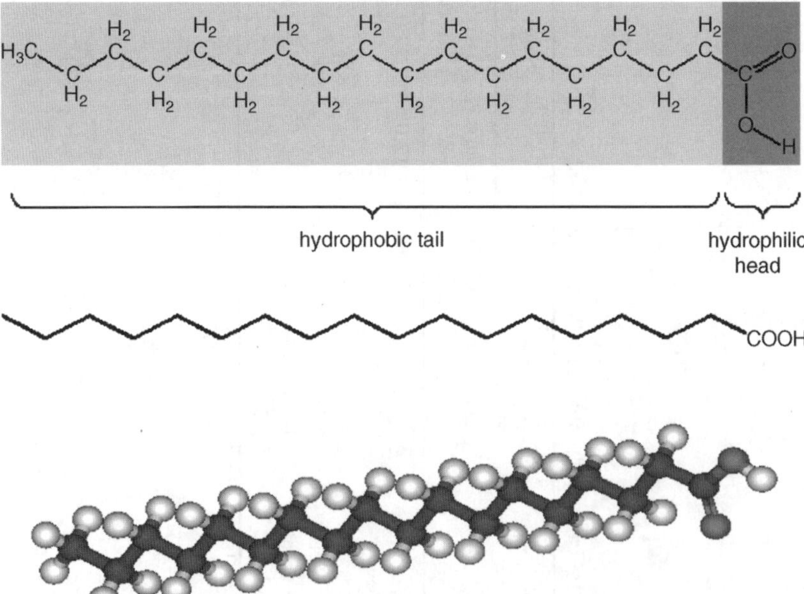

Figure 9.1 Stearic acid $C_{18}H_{36}O_2$ shown in three common formats. Courtesy of University of Oregon

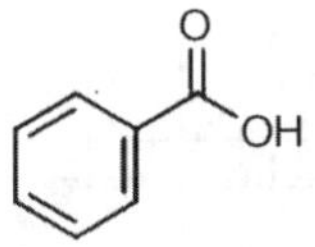

O
OH

Figure 9.2 Stearic acid $C_7H_6O_2$ (or C_6H_5COOH)

Another suitable fatty acid for soap making is benzoic acid, which also has a carboxyl (COOH) group, but instead of a straight chain like stearic acid, benzoic acid has an aromatic ring of six carbon atoms (Figure 9.2). It is the simplest aromatic carboxylic acid.

Selection of the base in soap or grease formation greatly affects the properties of the products. Common bases for saponification include sodium hydroxide (NaOH), potassium hydroxide (KOH), and most commonly lithium hydroxide (LiOH).

In making industrial soap, the [fatty] ACID of choice, in this case 12-hydroxystearic acid, is reacted (cooked) with lithium hydroxide (BASE). The acid and the base neutralize each other, or saponify, into a chemically neutralized soap. The process can be simplified as follows:

$$\text{FATTY ACID} + \text{BASE} \rightarrow \text{Soap} + \text{Water}$$

$$C_{17}H_{35}COOH + LiOH \rightarrow \text{Soap} + HOH$$

Lithium hydroxide (LiOH) is an alkali hydroxide that is corrosive. Commercially, it comes as a white hygroscopic crystalline material in anhydrous form, or as the monohydrate and is soluble in water. It is highly hydrophilic and care is needed to ensure the product does not absorb water before use and that its water content is known in the vendor-supplied certificate of analysis.

The water formed in the saponification reaction is removed from the soap by evaporation. After the soap is made, an equal amount of oil is mixed with it to make grease. Understandably, soap is a semi-solid and oil is a liquid, so the two will need to be mixed vigorously to create homogeneous finished grease. This is accomplished by using a homogenizer and/or a mill, which can impart high shears and ensures the oil to be thoroughly embedded within the matrix structure of the soap. To visualize, finished grease is like a sponge that is soaked with thick oil. The oil fills the cavities of the sponge and the sponge acts as a reservoir for the oil. If the homogenization is incomplete the oil could leak out, or, "bleed" out of the soap.

The process described above for making grease results in the formation of a "simple" lithium hydroxide grease. By introducing a SALT into the simple grease process, a soap COMPLEX can be formed with improved properties.

Making a complex soap (grease). When a SALT is introduced into a soap, this results in a product that is thicker than the original soap. When the complex thickener and the lubricating oil are mixed and homogenized, the resulting grease is a complex grease with improved properties over the simple grease. Different salts may be employed in the complexing of simple greases. For example, the complexing agent (salt) for the aluminum complex grease could be benzoic acid C_7H_6 O_2 or C_6H_5COOH. Benzoic acid is a simple aromatic (ring-structured) carboxylic acid, which is relatively a weak salt and has usage as a food preservative. This salt can be used as a complexing agent for other greases as well.

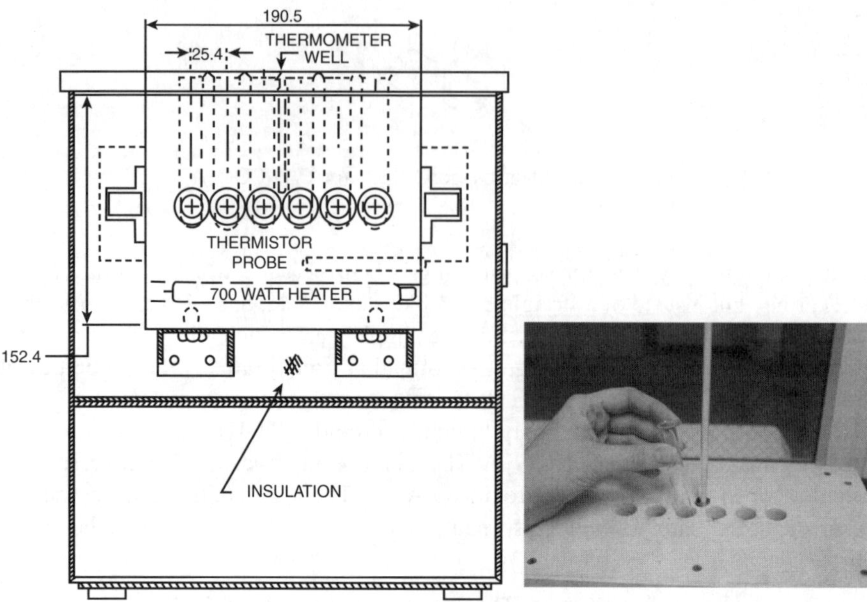

Figure 9.3 Schematic view of the Dropping Point Apparatus (left) and beaker and thermometer. Reprinted, with permission, from D2265-06 Standard Test Method for Dropping Point of Lubricating Grease over Wide Temperature Range, Copyright ASTM International, 100 Barr Harbor Drive, West Conshohocken, PA 19428. A copy of the complete standard may be obtained from ASTM International, www.astm.org

One of the most important benefits of complexing is the significant increase in the grease's ability to function at higher temperatures. A lithium complex (soap-salt) grease has a dropping (melting) point of about 50–60 °C (122–140 °F) higher than a simple lithium hydroxide grease.

The dropping point is determined by the use of a dropping point apparatus based on ASTM D 566 standard procedure. Figure 9.3 shows a dropping point apparatus where a thermometer monitors the temperature of the test sample while the sample is heated until the grease melts and drops, observed through the sight glass.

9.2 Basic Process for Manufacturing Grease

For soap-based greases, the majority have a similar manufacturing process and require reacting the base oil with a thickener at high temperatures, typically exceeding 138–149 °C (280.4–300.2 °F). For non-soap greases, like clay-thickened greases, room temperature reactions and high mechanical shearing can be used to manufacture the grease.

When lithium hydroxide is reacted with the base oil, a considerable amount of foam is generated along with water. This water will need to be boiled off at high temperatures, often exceeding 200 °C (392 °F). To remedy the foaming problem, high pressure sealed vessels may be used to contain the foam until the reaction is completed. The water will eventually have to be removed by venting the pressure vessel. Also, feeding lithium into the oil at a slower rate can help reduce the foaming, but it increases the process time.

Proper processing will ensure that the highest amount of grease is produced consistently from the same amount of thickener and the same quantity of oil. Unreacted lithium, failure to remove all the water, or overheating of the product will result in a myriad of variability issues, making the final product quality unpredictable. Hence, there is often a reference to the art aspect of grease making, whereby the operator relies on rules of thumb or judgment and gut feelings to adjust the process. Generally, grease manufacturers produce the grease at a higher consistency than desired, and then thin it back down by adding additional oil to reach the right consistency. This method works to some extent, but could also lead to excessive bleeding. Ideally, the grease manufacturing process should result in a consistent product at the right consistency/thickness without the need for thinning or mixing with thicker greases (reworking) to the right consistency.

In general, in any grease making process, the goal is to ensure the following:

- **Improved yield.** Yield refers to getting a higher consistency, thicker grease, with the same amount of thickeners, or more grease per unit of thickener.
- **Reduction in energy requirements.** Pressure via pressurized vessels may speed up the chemical reaction with less heat input and save energy.
- **Reduced production time.** Faster heating and reaction results in reduced production time.
- **Versatility.** To be able to heat and cool effectively and to control temperatures and other variables efficiently. Also, being able to switch from product to product and thickener to thickener will improve production and cycle time.
- **Operation savings.** This can be determined based on the cost per pound of producing the grease. Cumulative savings are considered in the overall grease operation.
- **Economical production with consistency and uniformity.** Any process that results in a high level of repeatability and consistent product quality between batches is desirable.

9.2.1 Simple (Soap-Based) Greases

Simple soap-based greases are made using the process of reacting the basic oil and fatty acid with the desired metal hydroxide to form soap. The processes explained here are very basic and require a considerable amount of experience to match the process with the manufacturing equipment and the process environment. Unfortunately, no single process can be standardized for all manufacturing environments and thus each manufacturing company would know their own process best due to their experience with their equipment. In the industry, this is referred to as the "art" rather than the science aspect of grease making. The National Lubricating Grease Institute (NLGI) has extensive, easy-to-follow publications that describe in more detail the manufacturing of these soap-based greases. They include:

(1) **Aluminum hydroxide soap-based grease.** Often pre-formed aluminum di-stearate soap is acquired and mixed with the oil at room temperature. The uniform mixture is then heated to 138–149 °C (280.4–300.2 °F) and mixed. This results in the formation of a gel-like substance, which is cooled in a separate vessel. Slow cooling results in a softer grease, while faster cooling would result in a firmer grease.

(2) **Calcium hydroxide soap-based grease.** For mineral oil-based greases, typically any variation of tallow and sometimes blends of tallow and fatty acid are used. The mineral oil, the fat [acid], and the alkali (in this case calcium hydroxide or hydrated lime) are mixed and heated to form the soap. Temperatures of about 150 °C (302 °F) are needed to complete the saponification, which is verified by testing the alkalinity of the product. Excess alkalinity, which is determined based on the process design, is neutralized through continued reaction until the saponification is complete. The completed soap is then mixed with the cooling oil (which is the lubricant portion of the grease). Due to its structure, the calcium soap does not

easily mix with the cooling oil at temperatures below 110 °C (230 °F). The addition of water to hydrate the soap has been shown to increase the soap's affinity to the oil and thus facilitate the mixing. The final grease may still contain up to 10% of the water content. As such, the grease is referred to as hydrated calcium grease.

If an anhydrous version of calcium grease is desired, then other processes are available to produce such a grease containing no water. Typically, 12-hydroxystearic acid is reacted with lime (calcium oxide) in the presence of the base oil to form soap, which is then mixed with oil to make grease.

Vegetable oils can be used as the base oil for either hydrated calcium soap or anhydrous calcium soap. The key is to first establish a consistent manufacturing process. Due to the presence of fatty acids in vegetable oils, only calcium hydroxide and anhydrous grease calcium oxide are reacted. The use of 12-hydroxystearic acid can still be helpful in the process.

New processes are also being introduced that use a fatty acid as a precursor to grease making. For example, Zhang [4] reported using oleic acid to create over-based calcium oleate complex greases. The complex reaction can be achieved through three different complex agent systems, each of which contains 12-hydroxystearic acid as one necessary component. It is believed that hydrogen bonding provided through the hydroxyl group on the resultant calcium 12-hydroxystearate makes a grease thickener system more robust and more shear resistant.

This approach is still based on oleic acid derived from either vegetable oils or animal fats, but it is more concentrated and has a higher total base number (330 g KOH/g) for increased grease thickening efficiency.

On the other hand, over-based calcium oleate grease precursors can be made with readily available raw materials. A schematic illustration of the manufacturing process is shown in Figure 9.4.

First CO_2 is introduced into a reaction kettle through a tube in a controlled manner in the carbonation process. This facilitates the formation of oil soluble calcium carbonate micelles, that is, nanoparticles of calcium carbonate dispersed in the oil phase with calcium oleate as a dispersant. Then larger particles, which are not oil soluble, are filtered out. Finally, solvent residues and other volatile polar components, which were necessary for the initial carbonation stage, are removed through vacuum stripping. A small amount of oil with a suitable viscosity is added to keep the finished grease precursor as a pourable liquid.

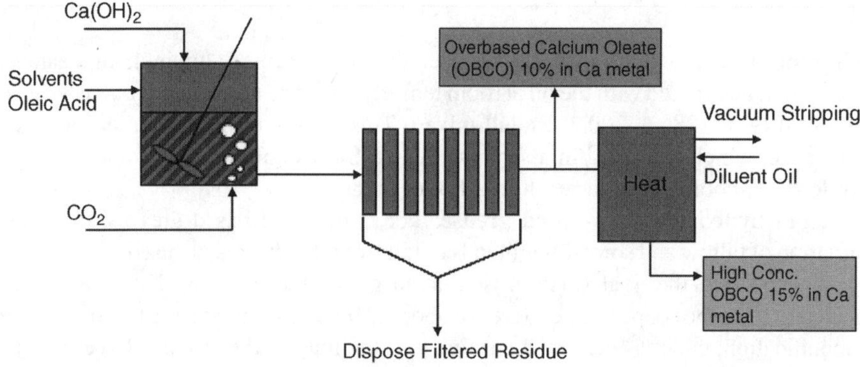

Figure 9.4 Manufacturing process for over-based calcium oleate

Table 9.2 Formulation for making over-based calcium oleate complex

Grease material	Mass (g)	Percent
White oil (600 SUS)	3250	56.00%
Grease precursor	2000	34.50%
Calcium hydroxide	100	1.70%
H_2O	350	6.00%
Propylene glycol	100	1.70%
Acetic acid	150	2.60%
12-hydroxystearic acid	200	3.50%
Total (excluding H_2O)	5800	100%

Zhang's example of a formulation for making such grease is given in Table 9.2.

By using white oil as the base oil, the final grease will also qualify as food-grade, based on accepted standards.

(1) **Sodium hydroxide (soda) soap-based grease.** The process for making sodium hydroxide grease is similar to other greases. The sodium hydroxide as a strong base is reacted with fats and fatty acid, in the presence of a small amount of water, to make soda soap. To avoid over-foaming, the heating of the reaction is slowed down and/or appropriate antifoams are incorporated. The soap in this process is fibrous and takes a high amount of energy to mix. Like with other greases, temperatures as high as 200 °C (392 °F) are sometimes incorporated to melt the soap thinner and make it easier to mix. This, of course, evaporates any water from the process as well. After completion of the saponification process, cooling oil is added and the soap-oil mixture is mixed and homogenized through milling.

(2) **Lithium hydroxide soap-based grease.** Lithium greases, both simple and complex, together make the largest portion of the grease produced worldwide. Figure 9.5 shows the estimated amount of each grease that is produced worldwide.

Similarly, the lithium hydroxide as a base is reacted with oil, fat, and fatty acid to make soap. Again some water may be added and the lithium hydroxide too may have absorbed water or been made into slurry to avoid exposure to its dust. The water is boiled off during the process. The initial process may require that the oil is heated to about 135 °C (275 °F) and then introduced to the lithium hydroxide. Excessive foaming is possible during the reaction. High-pressure vessels, anti-foam, or slowing of the reaction could be used to reduce or control foaming. A neutralized soap will have a pH around 7, indicating that all the base has reacted with the acid. After the reaction is completed, the soap is heated further to about 200 °C (392 °F) to evaporate the resulting water from the reaction. At this stage the cooling oil is added and the product is mixed, cooled, and milled until homogenized. When the product temperature is low enough to put in the chemical additives, they are then added to finish the product.

Lithium grease has a fibrous structure, which is effective in entraining the lubricating oil. Each of the greases described here has a different structure. Some, like the organo-clay grease, are very smooth and creamy with little fibrous structure. Figure 9.6 shows a scanning electron micrograph of a lithium hydroxystearate soap network after washing out the base oil [6].

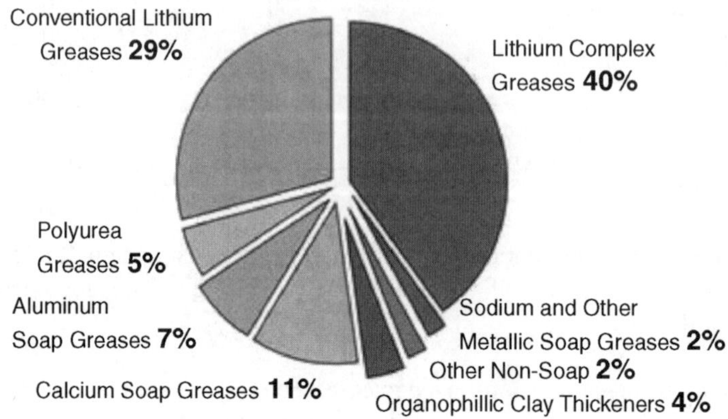

Figure 9.5 Grease production in North America based on thickener type [5]. Reproduced from the NLGI Spokesman, 2005, National Lubricating Grease Institute, Kansas City, MO, USA

9.2.2 Complex (Soap–Salt) based Greases

Complex greases are processed the same way as simple greases, except that they incorporate a salt as the complexing agent. Complex [soap–salt] greases include:

- Aluminum complex grease (aluminum hydroxide plus stearic acid and benzoic acid).
- Calcium complex grease (calcium stearate [soap] plus calcium acetate [salt]).
- Lithium complex (lithium hydroxide plus 12-hydroxystearic acid [soap] and dibasic acid or dimethyl azetate [salt]).
- Barium complex (barium and a fatty acid to make the soaps and use acetic acid as [salt] complexing agent).

9.2.3 Non-Soap-Based Greases

There are several types of greases made without the use of soap as their thickener. This is to overcome some of the limitations of various soap-thickened greases. Specifically, soap-based greases have limitations when exposed to extremely cold or hot temperatures. Aircraft greases, for example, are exposed to extremely cold temperatures at a high altitude. Chains on conveyors used in bakery ovens or foundry furnaces require greases with a high tolerance to heat. The presence of soap in grease makes the grease vulnerable to hardening at extremely cold temperatures or liquefying at extremely high temperatures.

Some of the non-soap greases require heat and reaction, similar to the soap-based greases. Others, however, use an insoluble material that is mixed with the oil to form a thickened lubricant with the same touch and feel of conventional greases.

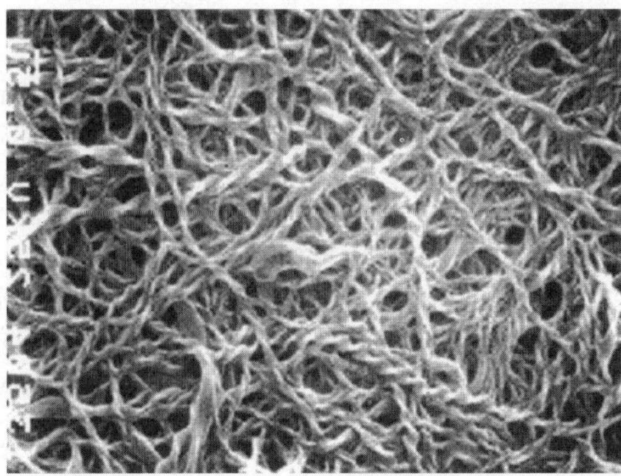

Figure 9.6 SEM image of lithium hydroxy stearate soap network without the base oil [6]

Polyurea and polyurea-complex greases are examples of non-soap greases requiring a chemical reaction. On the other hand, organoclay, carbon black, and silica are examples of non-soap based greases that are produced primarily through a mixing process. Other non-soap greases include polytetrafluoroethylene (PTFE), polyethylene, and polymer compounds. A brief description of these products follows.

Polyurea greases have the same consistency as soap-based greases, but they are not soap based and do not rely on saponification of fats or fatty acids. The lack of metals and acids make the polyurea greases oxidatively very stable. This is simply explained in the NLGI lubricating grease guide as a low molecular weight organic polymer. They still require a reaction of isocyantes and amines to form ureas. Both isocyanates and some amines are hazardous and require special handling during the processing. The reaction takes place at or near room temperature and no byproducts need to be removed. About a 7–12% thickener is used for most greases, and the grease generally has a very high dropping point of approximately 250 °C (480 °F), with excellent water resistance.

Polyurea complexes rely on complexing agents that improve some of their properties. For example, using calcium acetate or calcium phosphate results in a higher extreme pressure (EP) version of the simple polyurea version grease.

Organoclay thickeners are used for making grease near room temperature. Typically, the base oil is mixed with the organoclay, along with an activator like acetone, which can evaporate after the process is completed. Making clay-based grease requires high shear milling. A mixture of 8–12% clay and base oil are mixed and the activator is added. The mixture is then milled, resulting in thickened grease. Additives and performance enhancers can be added in a mixing vessel. Clay-based greases can be used in very high temperature applications like bakery ovens, as they typically do not have a dropping point. Vegetable oils can be thickened with organo-clay thickeners in the same way as mineral base oils.

Silica-based greases can be formed in a similar way to clay-based grease. Fume silica is added to the base oil and then milled to form the grease. Using esters and polyalphaolefins, silica-based thickeners are used as insulating materials for fiber-optic cables. For such applications the mixing is performed under vacuum to remove any entrained air. Silica-based greases do not have good resistance to water.

Other non-soap thickeners include carbon black, which in very fine form can thicken the oil effectively. Like clay-based greases, they have high temperature resistance but poor water wash out properties.

9.2.4 Preformed Soaps

To reduce the unpredictability in grease production, some suppliers have attempted to provide a pre-reacted dehydrated soap which will need only base oil and heat to prepare the grease. Other components like desired additives, coloring, and so on, are added to the mixture of soap and oil. Since the reaction uncertainty is taken out of the formula, the final product is more consistent and predictable. This is a highly desirable approach for smaller manufacturing operations and would eliminate the reworking of greases. This may also come at a higher cost for the final product due to the higher cost of the preformed soap. Vegetable oils can be effectively used as base oil with preformed soaps and could be a better approach than trying to react the oils with the thickener.

9.2.5 Preformed Dehydrated Soap for Biobased Greases

Dehydrated preformed soaps come in powder or flake form. They need only to be melted into the base oil to form the grease. Typically, temperatures as high as 200 °C (392 °F) are needed to melt the soap into the oil. Once the oil and the soap are melted, cooling oil is added to help with cooling and changing the soap into grease. The mixture thickens quickly and needs milling or homogenizing. Performance enhancing additives are added when the grease is cool enough. Figure 9.7 shows the preformed soap in powder form and the resulting grease made from melting the preformed soap in refined, bleached, and deodorized soybean oil.

9.2.6 Microparticle Dispersion of Lithium Hydroxide

Conventional lithium complex grease can be produced from either lithium hydroxide monohydrate solid or lithium hydroxide monohydrate water slurry. The process requires

Figure 9.7 Dehydrated preformed soap (left) and resulting grease (right). See Plate 28 for the color figure

Figure 9.8 Micro-particle dispersed lithium hydroxide is dehydrated solids dispersed in oil (left); grease made with vegetable oil base oil (right)

a significant amount of energy and generates foaming during saponification. In the case of lithium complex greases, a two-step process is frequently used in the conventional manufacturing method.

The Lubrizol Corporation (2009) introduced a new lithium hydroxide dispersion technology for making simple and lithium complex greases in a single-step process, promising reduced process time and improved efficiencies. By breaking the lithium hydroxide granules into fine, micron-sized particles, and mixing with oil, a larger surface area is created for reaction with the fatty acid and the liquid form makes it safer to handle (Figure 9.8). According to Zhu [7], in an open kettle the lithium hydroxide dispersion could save the grease manufacturer 40–60% in time compared to a conventional process.

Using more effective reaction vessels like those manufactured by US-based Stratco Inc., the use of higher pressure, and effective mixing can also reduce process time. In a Stratco Contactor™ Reactor, the process time savings from using lithium hydroxide dispersion are not as significant as that achieved in an open kettle; however, less energy is used by not having to heat and evaporate the added water.

In a continuous process one of the challenges is getting the right amount of lithium hydroxide into the reaction zone. Due to the poor water solubility of lithium hydroxide monohydrate, a large amount of water is needed to dissolve the lithium hydroxide. This requires the removal of significantly more water during dehydration. With the micron-sized microparticle dispersion of lithium hydroxide, the handling of the lithium becomes easier because there is no lithium dust, and the amount of foaming is reduced considerably.

9.2.7 Polymer-thickened Greases Using Bio-based Base Oil

The introduction of a polymer-thickened grease is somewhat new and is spearheaded by a Swedish company (Axel Christiernson, 2009). The polymer grease does not have many of the shortcomings of soap-based greases. Most importantly, since polymer thickener is non-polar, it does not compete with additives, most of which are highly polar. Figure 9.9 shows polymer grease made with rapeseed oil and illustrates the effectiveness of additives in the grease as compared to soap-based greases.

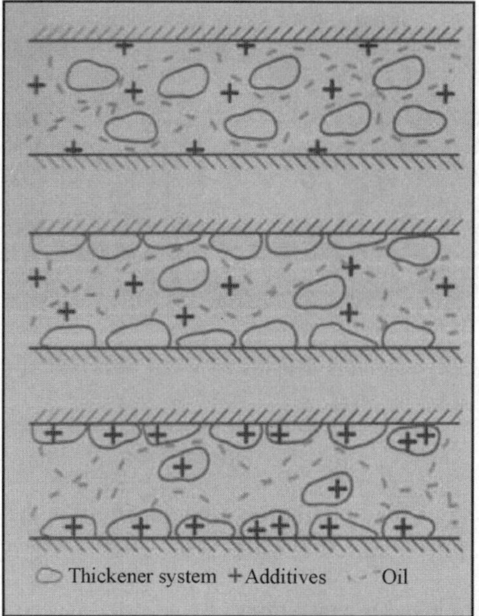

Epoch
Based on a non-polar thickener system (polypropylene) – the additives can reach the metal surfaces and do their job

Soap
Soap has a higher polarity than additives – many addtives will never reach the metal surface

Functional Soap
Soap and additives are bundled together – more additives will reach the metal surfaces, but some will inevitably be blocked in the middle.

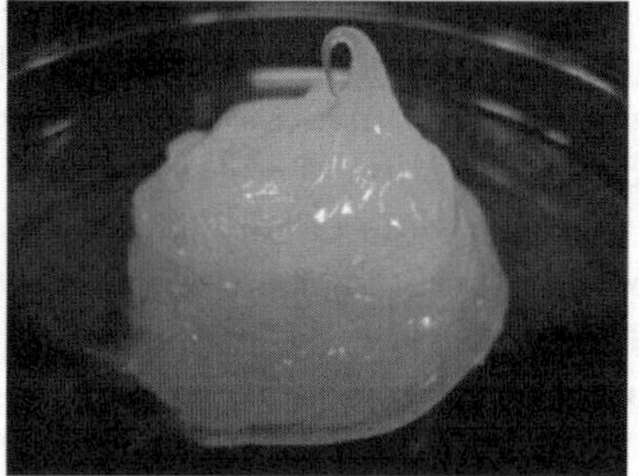

Figure 9.9 Polymer grease with rapeseed base oil and illustrating effectiveness of additives due to non-polar polymer thickener

9.3 Continuous Grease Manufacturing Process

Like the batch production of lithium and lithium complex greases, the continuous grease manufacturing process has been developed for both greases. Wittse *et al.* in a US patent [9] and Spagnoli *et al.* also in a US patent [10] describe the process for continuous complex grease manufacturing based on lithium complex thickener. In the continuous method, the saponification reaction product passes through a system that effects a pressure reduction and flash vaporizes the water from it. According to Spagnoli *et al.* [10], a mixture of dicarboxylic acids, a C12 to C24 hydroxy fatty acid in a mole ratio ranging from about 1 : 10 to 1 : 0.5, a lithium base, and a lubricating oil are continuously introduced into a reaction chamber. The base oil employed in making these greases may be conventional mineral oils and synthetic oils or blends thereof.

Accordingly, the mixture is heated to about 121 °C (250 °F) to 177 °C (350 °F). The reaction chamber is also held under turbulent mixing conditions, sufficient to obtain adequate contact between the reactants for a sufficient time to obtain a substantially complete reaction, forming a lithium complex soap. A product stream is continuously withdrawn from the reaction chamber. Then, additional lubricating oil is introduced into the product stream to give the grease mixture enough fluidity for circulation. The resulting grease mixture is continuously introduced into a dehydration chamber. During this process, a product stream is continuously withdrawn from the dehydration zone and cooled to provide a finished grease composition. This process, when designed properly, is expected to generate small quantities of finished grease on a continuous basis, thus resulting in a more consistent large volume production as compared to large batches of grease.

9.4 Use of High Pressure-High, Shear Reaction Chambers (Contactor)

In an attempt to create consistency in the final products, STRATCO, Inc. has coined the trademarked term Contactor to describe a unique, high-pressure, high-shear reaction chamber with benefits over open kettles. A Contactor is a smaller volume reaction chamber. It uses several features, including high pressure to subdue the resultant reaction foaming, high shear mixing to increase reaction efficiency, and full jackets to speed up the heating process. SRATCO, Inc. is a US supplier of Contactors and describes the "proprietary" Grease Contactor Reactor as consisting of a pressure vessel, a circulation tube, and a hydraulic head assembly with a mixing impeller and driver. A key feature of this reactor, according to the manufacturer, is the high turbulent circulation in a closed cycle path, which results in all the energy input through the mixing impeller to be expended within the materials being mixed. Figure 9.10

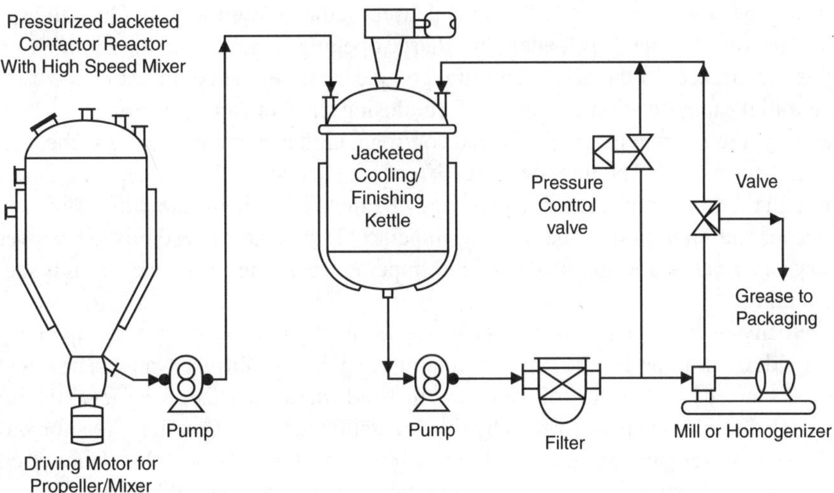

Figure 9.10 Process of making grease using a Contactor reactor. Source: http://www.stratcoinc.com/grease/ viewed January 10, 2011

shows a simplified illustration of the process using a Contactor reactor in conjunction with other needed equipment.

The process described here is the same for most grease manufacturing. Using a pressurized vessel to react the oil with lithium hydroxide helps in managing the foaming. Open kettles work the same way, except that the reaction is deliberately slowed down by a gradual feeding of the lithium hydroxide into the oil and by sometimes adding anti-foam agents into the mix.

In this process, the product is typically heated in the contactor reactor from 82.2–204.4 °C (180–400 °F) in 30–45 minutes with hot oil temperature at 232.2 °C (450 °F). For conventional kettles, hot oil temperatures would be around 260–287.8 °C (500–550 °F), but it would still take 3–5 hours for this same temperature rise. This is a claimed benefit of this style of Contactor reactor. Not only is the oil subjected to a lower wall temperature (at a higher velocity), but it is also subjected to the higher bulk reaction temperature for a dramatically shorter period of time. This results in a lower degradation of the base oil, especially useful for vegetable oils.

The STRATCO Contactor reactor uses high shear mixing to increase the efficiency of the reaction process. It consists of a pressure vessel, a circulation tube, and a hydraulic head assembly with a mixing impeller and driver. The outstanding feature of the Contactor reactor is its high turbulent circulation in a closed cycle path. All energy input through the mixing impeller is expended within the materials being mixed and there is virtually no pressure differential between the Contactor reactor's inlet and outlet.

The Contactor reactor is designed for an efficient saponification reaction in the manufacturing of grease. This equipment works well with biobased grease products because it reduces the process time and by efficient mixing reduces the exposure of the biobased oil to the hot surfaces of the reactor. The process as explained in the company literature is as follows.

The Contactor reactor is first charged with the specified base oil, which is at its ambient temperature. When the oil level is approximately 30% of the working volume, active circulation and heating are initiated. A hinged opening at the top of the Contactor reactor is used for charging the active dry chemical reactants used in the grease formula. Once the specified amount of the raw chemicals is loaded, the charge opening is closed and the Contactor reactor heating is maximized. In the case of lithium complex grease, the complexing agent is added after the initial saponification reaction before closing the Contactor reactor.

Using a hydraulic drive for ease of speed control, a high dispersion mixing in the Contactor reactor can be achieved. The impeller forces frequent changes in the velocity and direction of flow through the Contactor reactor. The oil and chemical ingredients are pulled down through the inside of the circulation tube via the impeller. Turbulence is established between the stationary shear vanes and the blades of the impeller, resulting in a zone of high shear and intense mixing.

The velocity of the material is greatly increased as it passes through the impeller. At the discharge side of the impeller, the stream is impinged against the diffuser vanes. The direction of the flow is reversed at the bottom of the hydraulic head where the vanes force an axial turbulent flow through the annular space formed by the circulation tube and Contactor reactor wall. The liquid then flows over the top of the circulation tube and back into the impeller. Once circulation is established, the Contactor reactor is heated, typically using a hot oil heating system.

The Contactor reactor is equipped with two jackets, an external jacket and an internal jacket. Heat transfer oil is circulated in both the internal and external heating jackets. High heat transfer coefficients are achieved by the high circulation rates across the heating surface. The saponification reaction is exothermic and the combination of applied heat and reaction heat results in a rapid temperature rise. The entire contents of the Contactor reactor reach a temperature in excess of 204.4 °C (400 °F) in less than an hour. When this temperature is reached, the saponification reaction is complete.

Pressures in the range of 70–100 psig (4.82–6.89 bar; 482–689 kPa) are typical of those used for efficient and essentially complete saponification. This pressure prevents foaming of the soap concentrate as well as accelerating the saponification reaction. The Contactor reactor pressure is maintained by manual or automatic venting as the temperature rises. As the saponification reaction proceeds to completion, water is generated as a product of the reaction.

Next the cooling (lubricating) oil is added to reduce the temperature below the soap crystallization temperature. At this point, the Contactor reactor's contents are pumped into a finishing kettle. These are large mixing vessels, typically a non-pressure or open atmospheric type. The kettles are equipped with large slow-speed mixing assemblies, whose function is to provide a gentle blending action and aid in cooling of the semi-finished product. Cooling is typically accomplished by cooling water circulation or cool oil flow through external jackets. Since most additives are temperature sensitive, they are added in the cooled kettles rather than in the Contactor reactor. Milling and finishing take place next as needed to complete the manufacturing of the product.

9.5 Vegetable Oil-based Greases

Vegetable oils can be used as a substitute for petroleum base oil in conventional grease manufacturing equipment. This means for simple lithium soap grease, the reactor could be charged with the chosen vegetable oil and heated to about 130 °C (266 °F), and then reacted with the lithium hydroxide. When the reaction is completed the resulting soap (having water as a byproduct) is heated while mixing to temperatures of around 204.4 °C (400 °F) to evaporate the water. Next, an approximate equal volume of cooling oil is introduced into the soap to make the grease, which is then milled to homogenize. The grease has to be cooled down to appropriately low enough temperatures so that the additives introduced in to the final product are not damaged or evaporated.

If the conventional grease making process requires fatty acids, because the presence of fatty acids in vegetable oils, the level of other fatty acids may be reduced. Nevertheless, the process yield is about the same as making grease with mineral oil-based greases. Also, vegetable oils may absorb atmospheric moisture, especially in summer. Thus, the grease manufacturing process may include more water that needs to be evaporated. The University of Northern Iowa's National Ag-Based Lubricants (UNI-NABL) Center has been engaged in the research and development of manufacturing processes and biobased greases made from vegetable oils. Several commercial grease products, including large volumes of rail curve grease made from soybean oils, owe their origin to this center. Since vegetable oils generally range in viscosity from 35 to 45 cSt at 40 °C (104 °F), UNI-NABL processes have included the introduction of some higher viscosity vegetable oils to increase the viscosity of the starting base oil. This could be the inclusion of blown vegetable oils or naturally higher viscosity oils mixed with the base oil. A high viscosity vegetable oil at 1500 cSt at 40 °C (104 °F) developed from an ionization process, produced by the Belgian company E'Ion, has been effective in increasing the viscosity of the base oil before reaction with the thickener.

Vegetable oils have a uniquely different behavior when exposed to high temperatures. In some vegetable oils, once the oil temperature exceeds 150 °C (302 °F), it begins to oxidize rapidly and if steps are not taken to remedy this rapid oxidation, the product will begin to polymerize, resulting in irreversible structural change. In such cases, the product could partially or fully polymerize or change its state from a soap into a polymer with little or no lubrication value. But, several methods exist for stabilizing soybean or other vegetable oils so they can be reacted with lithium hydroxide and produce stable greases. The use of high oleic vegetable oils is often used to improve the oxidation stability of the final product. Vegetable oils, due to their higher viscosity index, present a more stable body when exposed to high temperatures. As a result, properly formulated vegetable oil-based grease shows a more stable body in usage and will not thin down as fast as comparable mineral oil-based greases when exposed to high temperatures.

Vegetable oils present another challenge in grease making because they are made of mixtures of different fatty acids. Since different fatty acids have different melting points, when reacted with lithium, fatty acids with lower melting points react faster than fatty acids with higher melting points or longer carbon chains. It can be conceptualized that the vegetable oil-based grease can be considered a mixture of a number of different greases, such as lithium oleate, lithium linoleate, lithium stearate, lithium palmitate, and the like. These greases behave differently at different operating temperatures. The high oleic vegetable oils provide the advantage of making the final grease more uniform. For example, an ultra-high oleic sunflower oil with over 90% oleic acid would be mainly lithium oleate when reacted with lithium hydroxide.

9.5.1 Alternative Heating Methods [11]

A new process developed at the UNI-NABL Center has employed the use of microwave heating for the saponification reaction of the vegetable oils with lithium. This ground-breaking process has shown significant improvements in the process of grease making. In the laboratory and in production operations, it has been observed that the damage to the vegetable oils during reaction is primarily due to the methods of heating.

Current grease manufacturing processes include the use of heat transfer fluids being heated to high temperatures and then pumped into the heating jackets of the grease kettle. When the extremely hot heat transfer oil is pumped into the jackets of the reactor, the vegetable oil inside the reactor is exposed to the walls of the kettle, and thus, is exposed to very high temperatures via conduction through the metallic walls. It has been observed that when the vegetable oil is in contact with the walls of the reactor at temperatures of the kettle walls, it is immediately oxidized by a rapid chemical reaction. As hot materials rise, due to the convection process, additional layers of the oil take the place of the super-heated oil and the exposure continues, resulting in layer-by-layer oxidation of a portion of the base oil. Energy-wise, this method of heating is relatively inefficient because natural gas, fuel oil, or electricity is used to first heat the heat-transfer oil with some degree of inefficiency. The hot external surfaces of the pipes, kettles, and other units lose a considerable amount of heat even when insulated. The heat loss becomes excessive when the ΔT between heat transfer oil and ambient temperature is large.

Vegetable oils, due to their polarity, respond to microwave energy the way water does, and can be effectively heated by microwave with surprisingly high efficiency. When polar molecules of vegetable oils are exposed to high-energy microwaves, excitation in their molecules results in a rapid heat rise. It is theorized that the molecules of non-polar liquids, when exposed to microwaves, do not vibrate. Instead, they pick up speed and rotate, resulting in less friction at the molecular level and less heat rise.

Metaxas and Meredith [12] describe the polarization of a dielectric fluid as the interaction between an electric field with the dielectric as having its origin to the response of charge particles to the applied field. The displacement of charged particles from their equilibrium positions gives rise to induced dipoles, which respond to the applied field. As a result such induced polarization arises mainly from the displacement of atomic nuclei (electron polarization) or due to the displacement of atomic nuclei because of the unequal distribution of charge in molecule formation (atomic polarization) (p. 5). More factors that affect the polarization are involved but remain beyond the scope of this book.

The effective energy absorption of non-polar materials, like mineral oils, can be improved by the introduction of polar materials thus rendering them a better candidate for processing with microwave energy. But, the use of microwave energy for heating vegetable oils during grease processing is considered: (1) for avoiding degradation of oxidation stability due to exposure to high temperatures at the point of contact with the walls of the heating vessels; (2) to reduce the time needed to heat the oil to the reqiured reaction temperatures; (3) to reduce the energy consumption by a more focused and instantaneous energy input; and finally, (4) to reduce the level of fire hazard by eliminating the use of high temperature heat transfer oils. The following practical examples are provided to reinforce the theory behind the effective use of microwave energy for the manufacturing of biobased grease.

Table 9.3 and Figure 9.11 shows exposure of 300 ml of mineral oil and vegetable oil to 90 seconds of microwave energy through a 1.75 kW transmitter; then mixtures of mineral oil and vegetable oil and heating with the same level of 1.75 kW of microwave energy.

To simulate heating by conduction and convection, a sample of 300 ml of vegetable oil was placed on a hot plate and was heated to record the time needed to raise the temperature to 165 °C (329 °F). Similarly, the oil was exposed to microwave energy to reach 165 °C (329 °F). Figure 9.12, presents the results of this experiment, with noticeable differences in the time required.

Understandably the hot-plate method of heating will expose the heated oil to a longer period of heat loss from the walls of the beaker, and more accurate results need an adiabatic

Table 9.3 Microwave energy absorption*– mineral oil vs. vegetable oil

90 s microwave exposure sample	Temperature at start (°C)	Temperature after 90 s (°C)	Δ Temperature
HOBO**	22	109	87
Mineral oil	23	39	16
5/95 HOBO/mineral oil	23	44	21
10/90 HOBO/mineral oil	24	51	27
25/75 HOBO/mineral oil	24	60	36
50/50 HOBO/mineral oil	23	75	52
75/25 HOBO/mineral oil	23	96	73

*1.75 kW microwave input.
**HOBO – High Oleic Bean Oil.

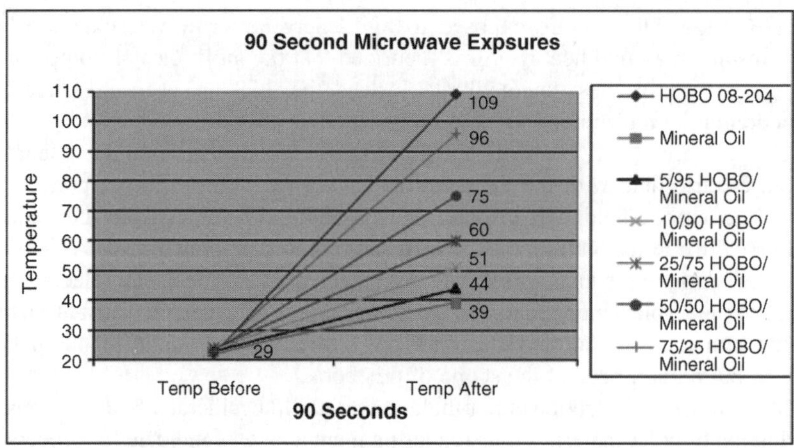

Figure 9.11 Absorption of microwave energy for mixture of vegetable oil and mineral oil (with different relative polarities)

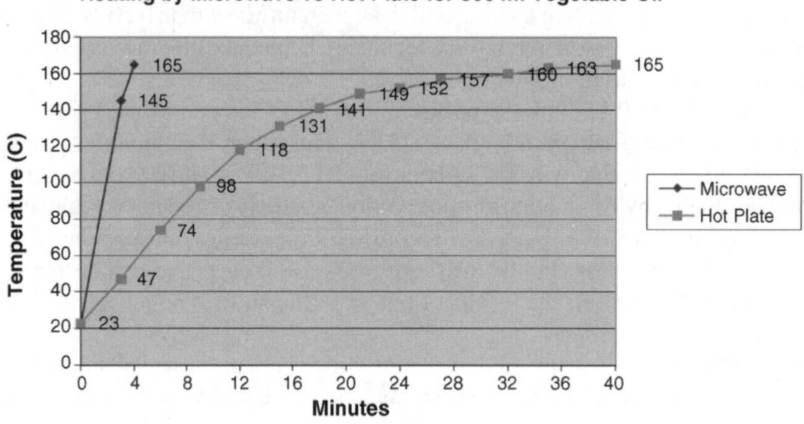

Figure 9.12 Time to raise temperature of vegetable oil to 165 °C (329 °F) by microwave and on a hot plate

environment. Nevertheless, this example shows the time-savings involved with the use of microwave energy when heating polar materials.

9.5.2 Heating Method and Impact on Oxidation Stability

To further investigate the effect of heating by microwave energy vs. conventional heating, a sample of vegetable oil with a known Oil Stability Index (OSI), was heated to 165 °C (329 °F) on the hot plate for 6 hours and a same size sample of the same oil was heated to 165 °C (329 °F) by microwave heating. The oil sample was then maintained at the same temperature by pulsing

Table 9.4 Impact of heating methods of oxidation degradation of vegetable oil

Vegetable oil RBD-HOBO[a]	Oil Stability Index (hours)
OSI before heating	41.2
OSI after heating – hot plate	9.12
OSI after heating – microwave	23.28

[a]Refined-bleached-deodorized high oleic bean oil.

one minute of microwave energy every 5 minutes for 6 hours (Table 9.4). The two oils were then tested for their OSI using an oxidation stability instrument. The results indicated that both oils oxidized due to extended exposure to heat, but the oil exposed to the heating on the hot plate had a change in its OSI of 2.5 times greater than that of the oil heated by microwave. In other words, the oil heated with microwave energy showed a 2.5 times better oxidation stability. Future reports will show that these trials can be duplicated in larger quantities with higher levels of microwave energy.

Further investigation in heating vegetable oils using microwaves indicated that the vegetable oil viscosity did not impact the absorption of microwave energy, as indicated by the increase in the temperature of the oils. A more detailed description of the factors impacting the absorption of microwave energy by various materials is beyond the scope of this book. The knowledge gained by investigating the use of a 1.75 kW microwave for heating vegetable oils in the laboratory was applied to larger volume oils in an industrial microwave system rated at 75 kW.

Figure 9.13 Vegetable oil and preformed soap heated with microwave energy to melt, and cooled to make grease

Figure 9.14 Microwave grease reactor in back with two 75 KW transmitters in front and waveguides for transferring microwave energy to the reactor. See Plate 29 for the color figure

The results indicated that the microwave energy would need to be matched to the volume of oil being heated. For example, using a 75 kW microwave system to heat a sample of 2 gallons (7.5 l) of soybean oil was not possible because two gallons did not provide a large enough mass for the large 75 kW microwave energy, resulting in sparking. Instead, the microwave energy level had to be reduced to 35 kW to avoid overwhelming the mass of the oil and to prevent heavy sparking. Figure 9.13 shows a 2-gallon (7.5 l) container of oil heated to 223 °C (433.4 °F) in 1.5 minutes, melting preformed lithium hydroxide-based soap to make grease.

The use of microwave energy for heating vegetable oils and processing greases is in the early stages of development. But, as the use of biobased products continues to increase in popularity, some deviation from conventional processes are expected. If uniform heating of the oil could be accomplished while preventing exposure to the high temperatures of the reactor walls, then more stable biobased greases can be manufactured. A production size microwave-based grease reactor using to 75KW microwave transmitter is shown in Figure 9.14.

9.6 Grease Consistency

In liquid lubricants the term viscosity is used to denote the oil's thickness or resistance to flow. In grease terminology, the term **consistency** refers to the thickness or thinness of the grease, which in effect is an indicator of the grease flowability.

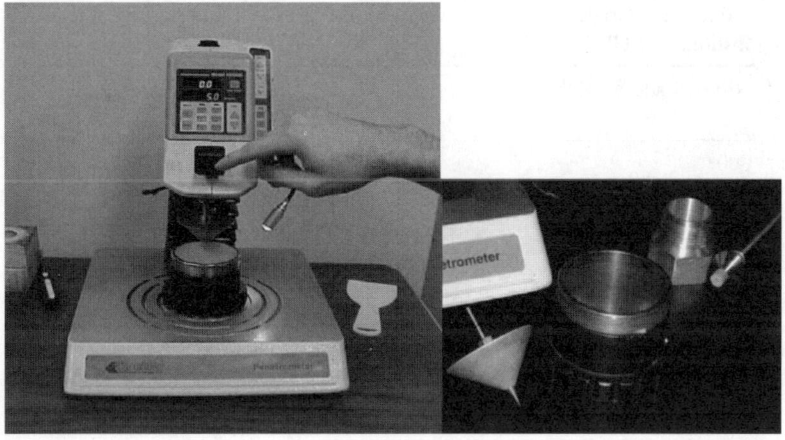

Figure 9.15 Grease penetrometer (left) and grease cups and standard cones (right)

Grease consistency is determined using the ASTM D 217 standard method, described later Based on theNLGI method of describing consistency, greases are identified by a numbering system, with lower numbers meaning thinner and higher numbers meaning thicker greases.

The consistency of a grease is measured with a device called a penetrometer (Figure 9.15). A standard cone-shaped object is dropped point first under controlled conditions into a sample of grease. The depth of penetration of the cone into the grease is measured in millimeters. This depth is referred to as the consistency of the grease.

Additionally, ASTM D 217 requires the grease to be "worked" utilizing a grease "worker" machine (Figure 9.16). The grade's for NLGI specifications are based on working the grease in a grease working machine for 60 strokes at 25 °C (77 °F). The grease worker has a disk with holes, which attached to the worker machine, would force the grease through the holes twice for every rotation of the grease worker arm. The working of grease or the forcing of the grease through those holes for extended periods of time simulates the environment in wheel bearings or other high shear applications. If the grease breaks down too much as observed by the penetrometer readings, then adjustment in additives may be needed. Grease workers are run for

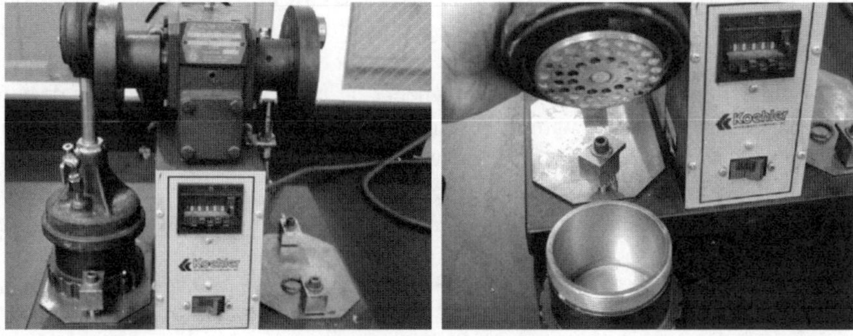

Figure 9.16 Grease worker (left) with grease standard cup

Table 9.5 Grade #s as designated by the National Lubricating Grease Institute (NLGI) [13]

NLGI Grade #	Penetration reading after 60 strokes at 25 °C (0.1 mm)
000	445–475
00	400–430
0	355–385
1	310–340
2	265–295
3	220–250
4	175–205
5	130–160
6	85–115

60 strokes to report the penetrometer reading. But, to better test long-term performance of greases, 10 000, 100 000, or even more strokes are also run and then tested in the penetrometer to determine changes. Greases could shear and thin down or, due to the breakdown of additives, break down and thicken up as determined by the penetrometer.

The resulting numbers are referred to as NLGI Grade Numbers and are expressed as NLGI Grade # 000 and up, as shown in Table 9.5. NLGI Grade #0, for example, is a thinner grease than an NLGI grade #2. For a grease to meet a grade number, the penetration value will need to fall within the range specified.

9.7 Grease Specifications

The NLGI in the United States is the primary body of experts focused on supporting the grease industry. Working with existing standards as developed by the ASTM, the NLGI continues to create new information useful in improving the performance standards of greases. Two other grease-related institutes, the India chapter of NLGI, and the European Lubricating Grease institute (ELGI), work with NLGI to ensure worldwide coverage of the needs of the grease industry.

Primarily due to the growth of the automotive industry and based on the request from the Society of Automotive Engineers (SAE), in 1989 the ASTM, with input from NLGI, approved ASTM D 4950 "Standard Classification and Specifications for Automotive Service Greases" (NLGI, 1996). ASTM D 4950 includes specifications for two categories of greases: (1) chassis lubricants (with letter designation L) and (2) wheel bearing lubricants (with letter designation G). Performance classifications within these categories result in two letter designations for chassis greases (LA and LB), and three for wheel-bearing greases (GA, GB, and GC). The automotive industry is in agreement that the highest performance classification in each group is LB and GC. This is typically shown as LB-GC, which is also suitable for service re-lubrication for automotive and truck shops.

Subsequently, NLGI developed the service categories for automotive greases to include: LA (chassis grease with frequent lubrication requirement), LB (chassis grease with infrequent lubrication requirement), GA (wheel bearing grease with frequent lubrication requirement), GB (wheel-bearing grease with less frequent re-lubrication), and GC (wheel-bearing grease for

Table 9.6 ASTM test methods required for ASTM D4950

ASTM D 212	Test method for cone penetration of lubricating greases
ASTM D 566	Test method for dropping point of lubricating greases
ASTM D 1264	Test method for water washout characteristics of lubricating greases
ASTM D 1742	Test method for oil separation from lubricating greases during storage
ASTM D 1743	Test method for corrosion prevention properties of lubricating greases
ASTM D 2265	Test method for dropping point of lubricating greases over wide temperature range
ASTM D 2266	Test method for wear preventive characteristics of lubricating greases (4-ball method)
ASTM D 3527	Test method for life performance automotive wheel bearing grease
ASTM D 4170	Test method for fretting wear protection by lubricating grease
ASTM D 4289	Test method for compatibility of lubricating grease with elastomers
ASTM D 4290	Test method for leakage tendencies of automomtive wheel bearing grease under accelerated conditions
ASTM D 4693	Test method for low temperature torque of grease lubricated wheel bearings

mild to severe duty) applications. A combination of these classifications is also used with the GC-LB as one of the highest classifications, normally used as a single grease that is suitable for all grease points of cars and trucks.

9.7.1 ASTM D4950 Specification

The ASTM D4950 is based on a set of 12 other ASTM tests as shown in Table 9.6.

The categories for each service class are shown as follows:

9.7.2 Service Category "L" Chassis (and Universal Joint) Grease

LA. Mild duty or non-critical applications. For use under highly loaded conditions with frequent relubrication (2000 miles or less). Table 9.7 shows the requirements for the service category LA.

LB. Mild to severe duty (high loads, vibration, exposure to water). Usable temperature of −40 °C to 120 °C (−40 °F to 248 °F) with re-lubrication intervals greater than 2000 miles. Table 9.8 shows the requirements for the service category LB.

9.7.3 Service Category "G" Wheel Bearing Grease

GA. Mild duty or non-critical applications. For use over limited temperature range with frequent re-lubrication. Table 9.9 shows the requirements for the service category GA.

GB. Mild to moderate duty (cars, trucks in urban and highway service). For use over a wide temperature range of −40–120 °C (−40–248 °F) and occasional spikes to 160 °C (320 °F). Table 9.10 shows the requirements for the service category GB.

GC. Mild to severe duty (frequent stop-and-go service, trailer hauling, mountain driving, etc.). For use over a wide temperature range of −40–120 °C (−40–320 °F) and occasional spikes to 160 °C (392 °F). Table 9.11 shows the requirements for the service category GC.

Table 9.7 Requirements for service category LA

Property	Requirement	Property	Requirement
Worked penetration	220–340	Elastomer compatibility	
Dropping point (°c)	80	3217/3B CR (%) volume change	0 to 40
Four Ball Wear (mm) max	0.9	3217/3B CR hardness change	−15 to 0
Four ball EP		3217/2B NBR-L volume change (%)	—
LWI (kg) min	—	3217/2B NBR-L hardness change	—
Weld Pt. (kg) min	—	Low temp. torque @ −40 °C (N-m)	—
Oil separation (%) max	—	Water resistance @ 80 °C (%) max	—
Rust (max)	—	High temp. life (h) min	—
Fretting protection (mg loss) max	—	Leakage tendency (g) max	—

Table 9.8 Requirements for service category LB

Property	Requirement	Property	Requirement
Worked penetration	220–340	Elastomer compatibility	
Dropping point (°C)	150	3217/3B CR (%) volume change	0 to 40
Four ball wear (mm) max	0.6	3217/3B CR hardness change	−15 to 0
Four ball EP		3217/2B NBR-L volume change (%)	—
LWI (kg) min	30	3217/2B NBR-L hardness change	—
Weld point (kg) min	200	Low Temp. torque @ −40 °C (N-m)	15.5
Oil separation (%) max	10	Water resistance @ 80 °C (%) max	—
Rust (max)	Pass	High temp. life (h) min	—
Fretting protection (mg loss) max	10	Leakage tendency (g) max	—

Table 9.9 Requirements for service category GA

Property	Requirement	Property	Requirement
Worked penetration	220–340	Elastomer compatibility	
Dropping point (°C)	80	3217/3B CR (%) volume change	—
Four ball wear (mm) max	—	3217/3B CR hardness change	—
Four ball EP		3217/2B NBR-L volume change (%)	—
LWI (kg) min	—	3217/2B NBR-L hardness change	—
Weld Pt. (kg) min	—	Low temp. torque @ −40 °C (N-m)	15.5
Oil separation (%) max	—	Water resistance @ 80 °C (%) max	—
Rust (max)	—	High temp. life (h) min	—
Fretting protection (mg loss) max	—	Leakage tendency (g) max	—

9.7.4 Multi-purpose Category

GC-LB. Multi-Purpose combined wheel bearing and chassis grease. There are, for example, soybean oil-based greases that meet these requirements. Table 9.12 shows the requirements for the service category GC-LB.

Table 9.10 Requirements for service category GB

Property	Requirement	Property	Requirement
Worked penetration	220–340	Elastomer compatibility	
Dropping point (°C)	175	3217/3B CR (%) volume change	—
Four ball wear (mm) max	0.9	3217/3B CR hardness change	—
Four ball EP		3217/2B NBR-L volume change (%)	−5 to 30
LWI (kg) min	—	3217/2B NBR-L hardness change	−15 to 2
Weld point (kg) min	—	Low Temp. Torque @ −40 °C (N-m)	15.5
Oil separation (%) max	10	Water resistance @ 80 °C (%) max	15
Rust (max)	Pass	High temp. life (h) min	40
Fretting protection (mg loss) max	—	Leakage tendency (g) max	24

Table 9.11 Requirements for service category GC

Property	Requirement	Property	Requirement
Worked penetration	220–340	Elastomer Compatibility	
Dropping point (°C)	220	3217/3B CR (%) volume change	—
Four ball wear (mm) max	0.9	3217/3B CR hardness change	—
Four ball EP		3217/2B NBR-L volume change (%)	−5 to 30
LWI (kg) min	30	3217/2B NBR-L hardness change	−15 to 2
Weld point (kg) min	200	Low temp. torque @ −40 ° C (N-m)	15.5
Oil separation (%) max	6	Water resistance @ 80 °C (%) max	15
Rust (max)	Pass	High temp. life (h) min	80
Fretting protection (mg loss) max	—	Leakage tendency (g) max	10

Table 9.12 Requirements for service category GC-LB

Property	Requirement	Property	Requirement
Worked penetration	220–340	Elastomer Compatibility	
Dropping point (°C)	220	3217/3B CR (%) volume change	0 to 40
Four ball wear (mm) max	0.6	3217/3B CR hardness change	−15 to 0
Four ball EP		3217/2B NBR-L volume change (%)	−5 to 30
LWI (kg) min	30	3217/2B NBR-L hardness change	−15 to 2
Weld point (kg) min	200	Low temp. torque @ −40 °C (N-m)	15.5
Oil separation (%) max	6	Water resistance @ 80 °C (%) max	15
Rust (max)	Pass	High temp. life (h) min	80
Fretting protection (mg loss) max	10	Leakage tendency (g) max	10

The NLGI symbols specifying the categories of grease are used for identifications by grease manufacturers under the NLGI license. Figure 9.17 shows examples of these symbols.

9.7.5 Dropping Point

Dropping point can be explained as a measure of grease's resistance to losing its property as a thickened lubricant. As the grease is heated it softens and loses its body thus running like a

Figure 9.17 Examples of NLGI labels licensed by manufacturers for labeling greases

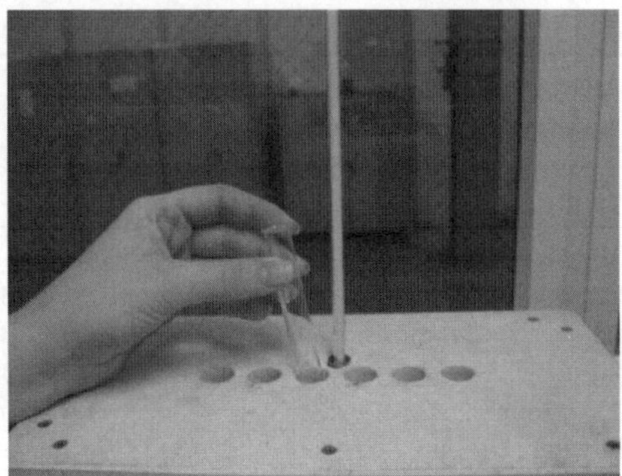

Figure 9.18 Dropping point apparatus

liquid. The test sample is a packed in a standard cup with a hole in the bottom. The sample is heated gradually while observing its temperature. The point where a drop of grease actually falls off the hole of the cup is considered the dropping point. Both ASTM D 566 and ASTM D 2265 are used to determine dropping points. ASTM D 2265 is a newer method that allows measurement over a wider range of temperature. A grease dropping point apparatus is shown in Figure 9.18. For the NLGI rated GC grease, for example, a minimum dropping point of 200 °C (392 °F) is needed for the grease to meet that requirement.

9.7.6 Water Washout

The water washout test is an older test that, due to its shortcomings, has been upgraded to a newer test. The old test ASTM D 1264 is still used and commonly specified, which is the ASTM "Standard Test Method for Determining the Water Washout Characteristics of Lubricating Greases." Simply put, the grease is packed in a tapered bearing while the bearing is run at 600 rpm, at 38 °C (100 °F) and water at about 5 cm^3/s (0.12 gallons per minute) is applied to the bearing housing (Figure 9.19). Some specifications may call for a higher water temperature of 70 °C (175 °F). The grease left in the bearing (subsequently, the grease that is not washed off the bearing) is measured to determine the grease's resistance to water washout. The problem with this test method is that the texture of the grease can impact the effect of water, creating a stringier grease to show a good water wash out, yet it would fail to perform in real world applications. Precision of this test especially suffers when smooth and creamy greases, like clay-based greases, are tested.

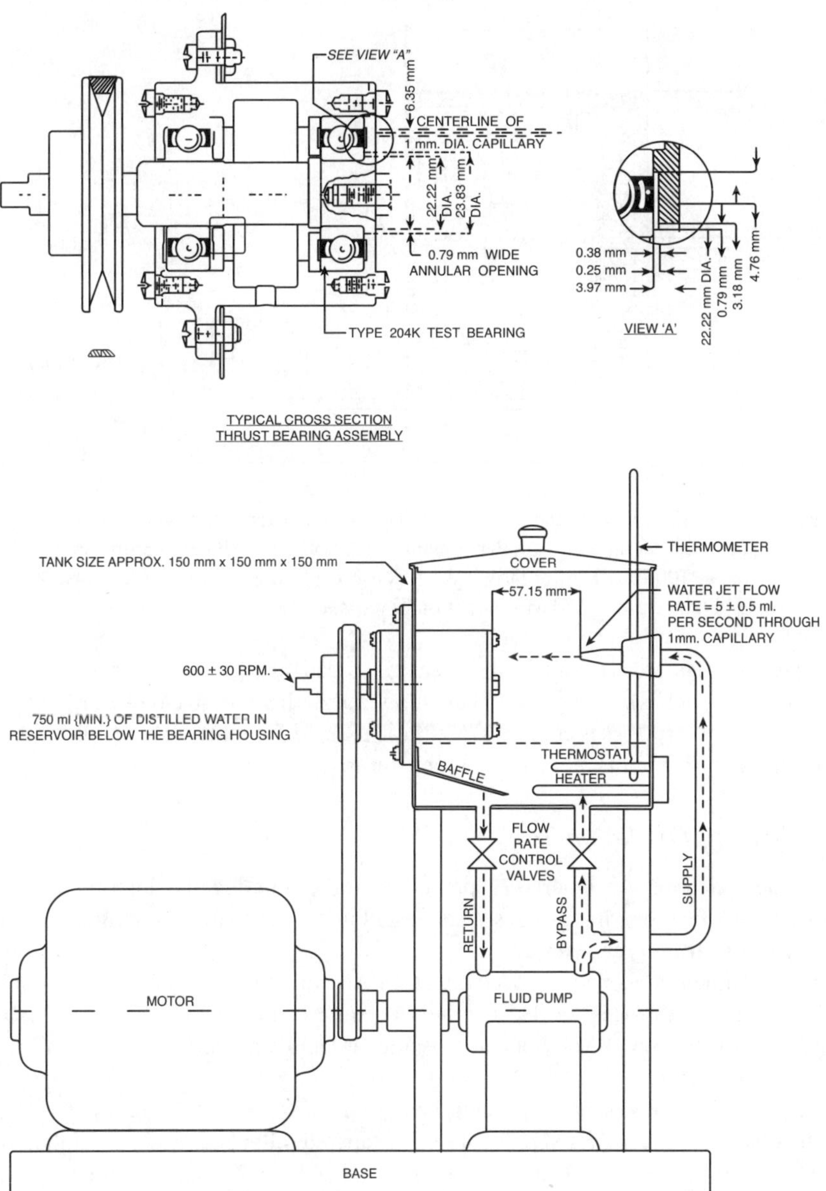

Figure 9.19 Water washout test apparatus for characteristics of lubricating grease. Source: ASTM International (2009), 100 Barr Harbor Drive, W. Conshohocken, PA 19428-2959, USA

9.7.7 Water Spray-Off

The water spray-off test uses a test panel and applies a spray of water to grease on the panel. The percentage of the weight of grease lost after a period of spray off is used to report as the test result. This test is not used for water-soluble greases.

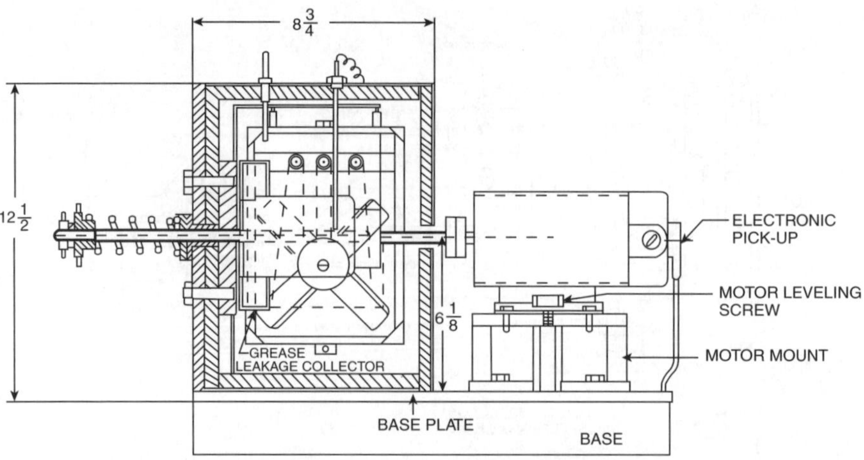

Figure 9.20 A schematic view of the wheel bearing oxidation tester. Reprinted, with permission, from D3527-07 Standard Test Method for Life Performance of Automotive Wheel Bearing Grease, Copyright ASTM International, 100 Barr Harbor Drive, West Conshohocken, PA 19428. A copy of the complete standard may be obtained from ASTM International, www.astm.org

The plates are standard size and the test grease is spread on the plate uniformly at a thickness of 0.8 mm (0.031 inches). Using a standard spray nozzle held at 30 cm (12 inches) from the plate, water at 38 °C (100 °F) is applied 275 kPa (40 psi), for 5 minutes. The weight of the grease left after the spray-off is calculated as the percent loss.

9.7.8 Bearing Oxidation Test

This test method covers a laboratory procedure for evaluating the high-temperature life performance of wheel bearing greases when tested under prescribed conditions. Figure 9.20 shows a wheel-bearing grease tester.

The test specimens are selected based on the user requirement or manufacturer's specification. The bearings and races needed for the procedure could be FLM67048-LM67010 and FLM11949-LM11910 (AFBMA Standard 19) inboard and outboard bearings, available from equipment manufacturers.

The new tapered roller test bearings come coated with an antirust coating which is cleaned in a suitable solvent as per the ASTM procedure. The tapered roller bearings are weighed and then the races are installed into the cleaned hub of the test machine. The test bearing is then packed with the test grease. The test involves monitoring the starting torque of the motor when starting and then running the bearing at a speed of 1000 ± 50 rpm, for 20 hours, and then letting it sit for 4 hours. After the fourth hour, the running torque is again monitored for the next 20 hours. The motor cut-off current value in amperes, the steady-state current in amperes, and the unloaded motor current in amperes are used to determine the pass/fail of the grease. There is an automatic torque cut-off that can be set to 10.8 for ASTM D 3527 and 10.10 for ASTM D 4290. As the grease is exposed to high heat during each 20-hour run, it is oxidized and its consistency increases (thickens), requiring more torque to run in consecutive runs. If the grease thickens too much, the machine shuts down when the automatic torque cut-off value is reached.

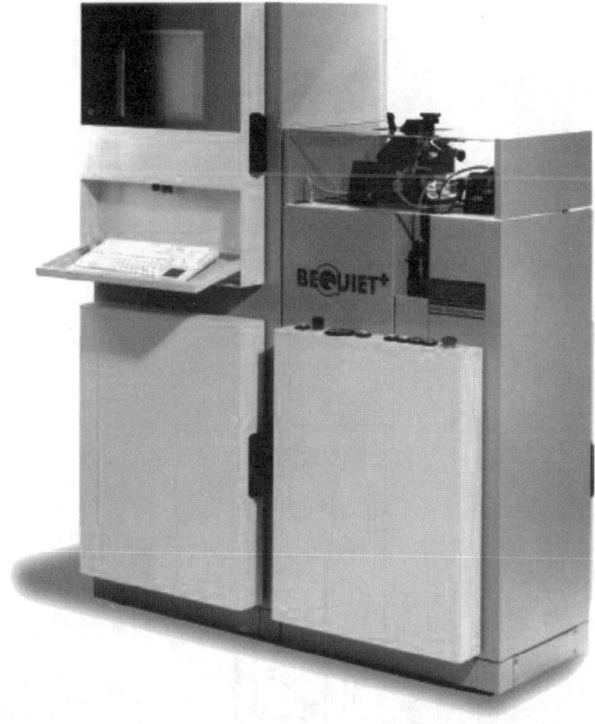

Figure 9.21 Bearing noise testing. Courtesy of SKF

This test has many variables and may require more than one test to ensure that failure or success is not due to variability of the test or the test equipment. Vegetable oil-based greases do not perform well in this test, which is required to meet the GC-LB (explained later in this chapter) requirements. But some commercial vegetable oil-based greases are on the market that meet the GC-LB and the requirement of this test. Figure 9.20 shows a bearing oxidation tester.

9.7.9 Grease Cleanliness and Noise

In the case of grease many factors can affect the degree of cleanliness during operation. But, manufacturers of grease also filter the grease to meet various required cleanliness. Also in applications where the bearing fatigue life is not critical (like very low loads), the need for clean greases can still be extremely important as they contribute significantly to low bearing noise, which is an important requirement for certain applications like electric motors in appliances or computers.

Greased bearings can be tested for noise level using different types of equipment. One test unit, developed by the SKF company, tests for the noise level of bearings under various microloads (Figure 9.21).

9.7.10 Grease Mobility Test

The reference method for this test comes from the US Steel Method: "Determination of grease mobility at cold temperatures." There is standard equipment on the market for performing this

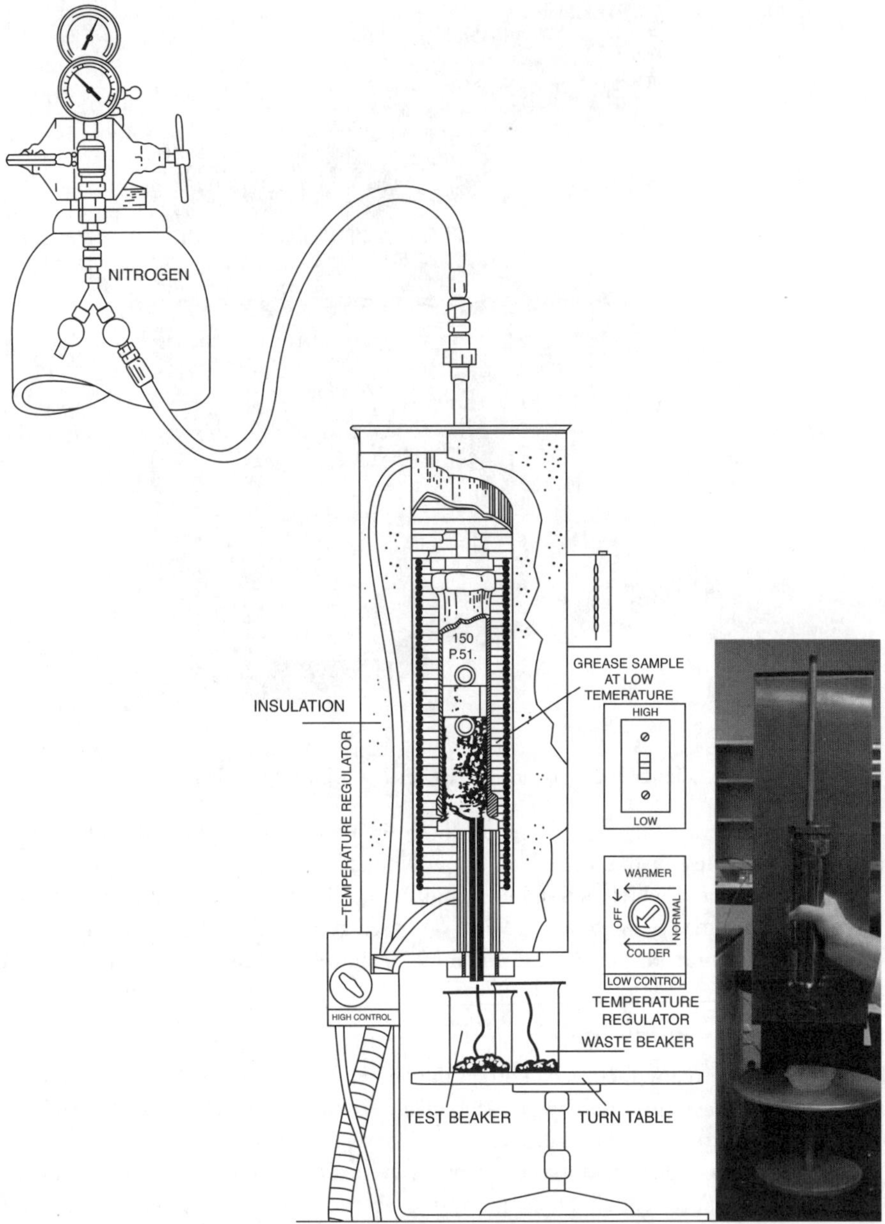

Figure 9.22 Illustration of the grease mobility tester (left). Source: Kohler Instruments: http://www.koehlerinstrument.com/ and grease chamber removed (right)

test, shown in Figure 9.22. In this concept, the test requires a cylinder with holes on both ends that can hold the test grease and be cooled to desired temperatures. Then, 150 psi (103 kPa) of nitrogen pressure is applied to the chamber for a specified time, until 50 mg of grease is pumped out of the cylinder.

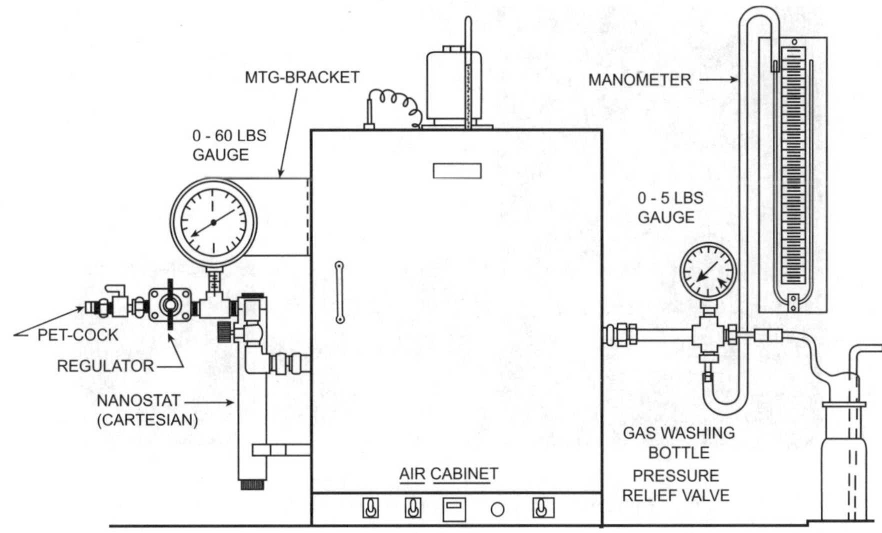

Figure 9.23 Illustration of the test set up for oil separation from lubricating grease during storage. Source: Kohler Instrument: http://www.koehlerinstrument.com/

From the weight of the grease pumped and the time it takes to get 50 g of grease, a flow rate in grams per second is calculated. This test is especially important for greases that are used on truck automatic grease systems where long steel tubes deliver the grease to various undercarriage greasing points of the truck. The flow rate at different temperatures can be reported and matched against the equipment manufacturer's recommended flow rates.

Sometimes other factors such as inadequate milling or homogenization of the soap and oil could result in the oil separation. To determine the propensity of the grease to bleed out, the ASTM D 1742 is used (Figure 9.23). This test is performed at room temperature which may not be adequate for the applications where the grease is exposed to higher temperatures. In such cases another test – Methods 321 of the Federal Test Method Document 791 – can be used. In this test, a wire screen cone made up of 60-mesh screen would contain the grease is suspended in a beaker. The beaker is then kept in an oven at 100 °C (212 °F) for 50 hours. The bleed out is measured form the quantity of the oil in the beaker.

9.7.11 Evaporation

In some applications especially when high temperatures are involved evaporation of the oil in the grease could become a problem. Biobased greases in general show lower evaporation loss due to the lower volatility of the base oil. But, biobased grease manufacturers may include low volatility oils or additives and sometime thinner synthetic oils to help with cold temperature flowability. In such cases, the presence of the low viscosity oil or low volatility oils or additives contribute to excessive evaporation loss. To determine evaporation loss of greases, the ASTM D 972 and ASTM D 2595 (for higher temperatures) are used.

The test method involves holding the grease in a chamber heated to 99–149 °C (200–300 °F) and blowing preheated air at 2 l/min over the grease surface for 22 hours (Figure 9.24). The

Figure 9.24 Rotary bomb oxidation tester

evaporation loss is determined by the difference between the weight of the grease before and after the test. The ASM D 2595 requires a higher temperature for the chamber and the air; 93–316 °C (200–600 °F). When vegetable oil only is in the biobased soap (grease), the evaporation loss is reduced significantly. As mentioned earlier, the presence of volatile additives or thin oils to help with cold temperature flowability could result in increased evaporation loss in biobased greases.

9.7.12 Oxidation Stability for Storage of Biobased Greases

There is no universally accepted test of oxidation that could relate to stability in storage. This is especially a problem in military applications where the shelf-life requirements are more stringent than for civilian applications. Also, vegetable oils and vegetable oil-based greases have traditionally not performed well in oxidation tests designed for petroleum-based oils or greases.

A test that is often used for oxidation and by default for storage stability is the Bomb Oxidation Test (ASTM D 492), which was designed as a *service* oxidation test (Figure 9.24). In this test, five glass dishes each holding 4 g of the test grease are placed in a rack inside a sealed chamber (bomb) and charged with oxygen at 758 kPa (110 psi) and at 99 °C (210 °F). As oxygen is absorbed in the grease – oxidation – the pressure in the bomb drops. A pressure drop of *X* kPa at *Y* hours is then expressed as oxidation stability of the test grease. For biobased greases, the emphasis is on placed on the oxidation stability of the cooling oil which is mixed in the soap after the soap is formed in the manufacturing process.

Since the reaction of the fatty acid (in this case vegetable oil) with a *base* (e.g. lithium hydroxide monohydrate) creates a soap, here it is assumed that full structural change has taken place. In this case, the stability of the soap is different from the oxidation stability of the original fatty acid. But the oil that is entrained in the soap, as lubricating oil, is the focus of the test for oxidation. Using the Oil Stability Index (AOCS Method Cd 1 2b-92), and selecting an oil with a high oxidation stability number should result in a high stability biobased grease. Note to self – cite NLGI book as reference.

9.7.13 Oxidation Stability in Service

The concept of oxidation stability for service for biobased greases is the same as oxidation for storage as described above. But, for petroleum products, in addition to the ASTM D 1741 bearing test – described earlier – the ASTM D 3336 is used. In this case, a smaller bearing size 204 is run at 10 000 rpm and at a temperature of 371 °C (700 °F). The test is run by cycling 20 hours on and 4 hours off cycle with a load of 5–15 lbf (22–67 N) until failure.

9.8 Friction and Wear Tests

9.8.1 Four-ball Wear Test and Four-ball EP

Weld point and load wear index are often used for determining the properties of biobased greases. For some rail curve applications, for example, EP properties are important because of the heavy side loads exerted by the wheel flange against the track. EP ratings of 500 kg or more are specified for some rail curve greases. Table 9.13 shows typical properties of a commercially available soybean oil-based rail curve grease.

Other tests include wheel bearing leakage, roller stability, evaporation loss, oil separation, and storage stability.

9.9 Application Examples of Biobased Greases

9.9.1 Rail Curve Greases

Many vegetable oil-based greases have been successfully commercialized in the United States. The most well-known product was created at the UNI-NABL Center for use as rail curve grease. Commercialized under a license, the grease was matched to the operation of a grease dispenser, manufactured by Portec Rail Products, Inc. The grease dispenser, also referred to as a lubricator, has advanced to include automation and unsupervised operation.

Table 9.13 Properties of a commercial soybean oil-based rail curve grease

Rail curve grease property	Value
Appearance	Dark green to black
Cone penetration, unworked	310–340
Four-ball wear scar (mm)	0.58
Four-ball weld load (kg)	500
Dropping point °C (°F)	191 (376)
Base oil viscosity at 40 °C (cSt)	86
Base oil viscosity at 100 °C (cSt)	16
Base oil viscosity index	201
Base oil flash point, °C (°F)	326 (619)
Base oil biodegradability	>60% in 28 days
Base oil aquatic toxicity	Non-toxic

Figure 9.25 Hi-rail grease applicator installed on hi-rail truck driving on the track (On-Board lubricator web site)

On-board grease lubricators that use thinner grease and apply the grease directly to the flange of the locomotive wheels can be employed. Some smaller railroads use "hi-rail" lubrication methods where a maintenance truck with flanged wheels drives on the track and applies a bead of grease to the inside face (gage face) of each track (Figure 9.25).

Other lubricators sit by the side of the track and are equipped to sense the presence of the train wheels and apply the grease as needed. Figure 9.26 shows the operation of an

Figure 9.26 Wayside automated rail curve grease lubricator. See Plate 30 for the color plate

automated lubricator. The lubricator has a reservoir that can hold up to 200 gallons (750 l) of grease, with a positive displacement pump and appropriate valve and plumbing to get the grease to the dispensing "bars" that are attached to the gage face (inside face) of the tracks. The pump can be run by a hydraulic motor powered by a mechanically operated plunger, which is activated by the flange of the passing wheels of the train as they go by; or by an electric motor driven by a battery. The battery-operated units are charged by an array of photovoltaic solar panels.

For the battery-operated units, a proximity sensor, placed at the side of the track before a curve, senses the presence of each wheel as the wheels go by. The sensor's signal can be programmed to turn the electric motor on for a controllable period of time. For example, if for each passing wheel the pump is set to run for 0.5 s, then for every two wheels passing the sensor, the pump would keep running for 1 s. These adjustments are accurate and can be set to deliver a sufficient amount of grease and avoid under-greasing or over-greasing. Under-greasing does not allow the wheel to carry the grease to far enough distances through the curve, while over-greasing can cause the grease migration to the top of the rail which us undesirable and can cause slippage of the wheels.

Vegetable oil-based rail curve greases have successfully been employed and seem to be a good lost-in-use type application. But, the product formulation can be complicated, as the grease requires the right amount of tackiness to adhere to the wheel flange (required adhesion). However, the grease cannot be too cohesive, as to prevent swing around the wheel and building up under the railcars (required cohesion). Also, the grease is exposed to varying temperatures and needs to have a stable body in changing temperatures to ensure uniform quantities are delivered for each wheel.

Figure 9.27 shows the components of an automated wayside rail curve grease lubricator.

The effectiveness of greasing the track is determined by the use of a tribometer, which can be handheld and operated manually or can be truck mounted for high-speed measurements. An instrumented wheel is run against the gage face or the top of the rail to measure the available coefficient of friction. A coefficient of friction of 0.23–0.28 would be desired for the gage face. Figure 9.28 shows a hand-operated tribometer used to monitor the coefficient of friction on a rail track.

Figure 9.29 shows illustrations of the concepts used in determining the coefficient of friction using a track gage tribometer as well as a hi-rail based high-speed tribometer.

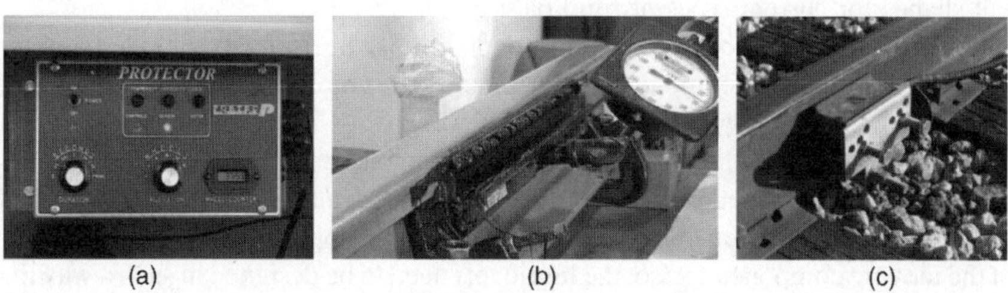

(a) (b) (c)

Figure 9.27 Wayside grease lubricator components: (a) controller, (b) a dispenser bar, (c) proximity sensor

Figure 9.28 Operation of hand-held tribometer. See Plate 31 for the color figure

Figure 9.29 Operation of hi-rail high speed tribometer (Hi speed tribometer from Portec)

9.9.2 Solid Lubricants

Biobased oil-based solid stick lubricants [14] – rigid-structured lubricants are commonly used in applications where the use of liquid lubricants and oil reservoirs or other components is not practical or feasible. These rigid lubricants, also called solid or stick lubricants, can be preformed into any shape to adapt to the components being lubricated. They are often made of a lubricating component such as graphite or molybdenum disulfide, and formed into a solid shape by the use of various adhesives, polymers, or other bonding agents. Figure 9.30a– shows four different shapes for current types of rigid or stick lubricants.

Conventional solid lubricants used for locomotive wheels have included mixtures of polymers and lubricating components such as graphite, molybdenum disulfide, Teflon, or mineral oil.

The benefit of rigid-structure lubricants is, when compared with liquid lubricants, that they do not require lubricant reservoirs, lines, or other components. Rigid-structure lubricants do require special applicators or dispensers to continuously apply force to press the lubricant against the moving component. Also, the lubricants need to be designed in such a way that they do not melt too quickly at high speeds of the moving component and so that they wear at a controlled rate. In fact, controlling the rate of wear is an extremely important criterion in the success of solid lubricants.

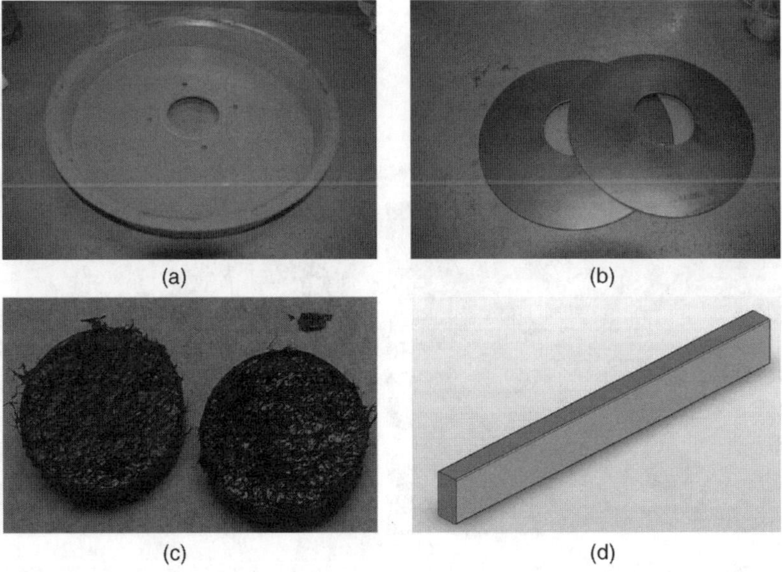

Figure 9.30 Solid lubricants shaped for intended application

While an important and effective lubrication technique, solid lubricants do not replace rail curve grease and other lubricants for track lubrication. Solid lubricants do, however, provide suitable lubrication for the flanged wheels of locomotives, thus improving wheel and track longevity, and reducing energy use by reducing drag. At the same time they reduce potential top of rail contamination. Other applications could include chains on conveyor systems, ropes used in shipyards, ski lifts, escalators, and so on.

Using lubrication of locomotive wheels, the use of solid lubricants is exemplified. Locomotive stick lubricants are useful because they can be applied directly to the wheel flange. This eliminates the possibility of the lubricant migrating to the flat side of the wheel, or to the top of rail. Alternate methods of lubrication, like grease or onboard lube sprayers, often result in the migration of the lubricant to the top of rail, resulting in wheel slippage and braking problems.

A solid stick lubricant dispenser, as shown in Figure 9.31, uses a constant pressure spring (Figure 9.32) to apply a constant force to the lubricant stick. The lubricant, in turn, is pressed against the flange of the locomotive wheel where it constantly applies the desired amount of lubrication.

These dispensers are designed for ease of operation and require some type of auxiliary force, which is typically applied by a constant force spring. Figure 9.31 shows a rolled spring which is designed to handle the vibration and changing temperatures, yet provide a constant application force throughout the length of the lubricant stick.

Conventional solid stick lubricants are 43–45 cm (17–18 inches) in length, 2.2 cm (1 inch) thick and about 5.5–6.6 cm (2 1/2–3 inches) wide. They are manufactured under various heat and pressure conditions. Some are extruded into shape, while others may be pressed into a form.

Figure 9.31 Solid stick lubricant dispensers: (right) courtesy of Snyder Company, (left) courtesy of Kelsan Technologies Corp. (http://www.kelsan.com/)

Figure 9.32 Dispenser (right), rolled spring positioned on the lubricant stick (left)

The density of the material as well as the type of bonding used determines both the degree of rigidity and the rate of wear.

Because the dispensing unit is placed 5–7.5 cm (2–3 inches) away from the face of the wheel, the last 5–7.5 cm (2–3 inches) of the stick lubricant falls out of the dispenser unused. See Figure 9.33.

The unused portion of the stick lubricant becomes a wasted material that can also be a source of pollution. In rare cases, left-over pieces could fall within the switching mechanism and cause safety hazards.

Figure 9.33 Residual piece of stick lubricant

Figure 9.34 Interlocking design. Courtesy Kelsan Technologies Corp.

Some manufacturers have incorporated interlocking designs that allow maintenance technicians to attach a half-used stick to a new stick, thus eliminating waste. Figure 9.34 shows one design where each stick lubricant is interlocked with the next one allowing continuing addition of sticks, and preventing pieces from falling out of the dispenser.

When there is a need to deliver liquid lubricant from a rigid-structure lubricant stick, the lattice can be made of thermoplastic or thermoset materials that are compatible with the desired lubricant. Figure 9.35 shows an experimental stick lubricant with a lattice containing liquid lubricant.

This is by far one of the most effective ways to deliver small amounts of liquid lubricants over an extended period of time. The size of the cells determines what the viscosity of the liquid lubricant can be. If the lubricant has low viscosity, then the cell size would be small. For comparative purposes, cell sizes could be compared to the mesh size of screens. A 20 mesh would contain 20 cells per inch versus a larger mesh like the one in Figure 9.35 where 4 cells per inch were used to enclose liquid lubricants with the consistency of gear lube. When a solid lubricant can release the fluid at a controlled rate it, in effect, acts as a complete system containing reservoir, piping and control valves.

Figure 9.35 Solid stick lubricant, lattice containing liquid lubricant (top) and hardened grease (bottom)

A simpler, yet equally effective method of controlling the rate of stick lubricant wear is the inclusion of a mixture of natural fiber materials into the mixture of the lubricant and biobased polymers. The fibrous materials soak the oil from the grease, and act as a wick, while helping to retard the rate of wear. This method, while easier to manufacture, does not provide the cell structure used to store the lubricant.

In the case of solid stick lubricants made with soy-based grease, the wear rate can also be controlled by inclusion of "wear bars," which are longer-wearing polymers placed along the length of the stick. This method allows the use of a mixture of the fibrous material, the lubricant, and the polymers in one mixture. The amount of fiber and the density of the polymer can be used to increase or decrease the wear rate. But, inclusion of wear bars provides more flexibility in controlling the wear rate. This approach results in an easier manufacturing process by removing the steps needed to include the lattice. Figure 9.36 illustrates how the wear bars are placed within the stick lubricant.

9.9.3 Truck Greases

Truck greases present another suitable application for biobased greases. Most of the grease is typically melted away or washed off by rain or in truck wash, ending up in the environment. The simplest application is the hitch point of the tractor, also called the fifth wheel.

Biobased greases that meet the GC-LB requirements of NLGI are now available commercially and can be used in all automotive components that need greasing. When grease is to be used in centralized automatic greasing systems, the cold temperature flowability becomes increasingly critical. Such greases are often thinner to ensure flowability in cold temperatures. Figure 9.37 show a truck fifth wheel, which requires greasing to ensure ease of handling, especially when cornering.

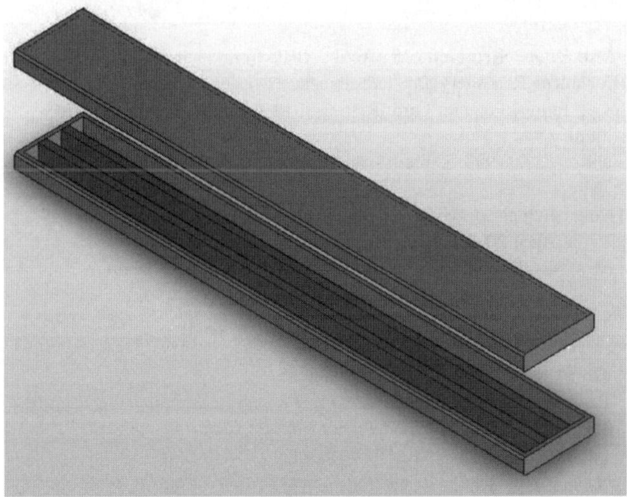

Figure 9.36 Placement of two wear bars along the length of lubricant

The biobased grease can also be formulated in a way to be useable in trucks with centralized greasing systems. Such greases are thinner than regular truck greases and would require testing of cold temperature flowability to ensure they flow at cold temperatures.

Cotton picker spindle grease, Food grade machinery grease, recreational paddle-boats, and open gears are examples of other applications where biobased greases can provide superior performance.

Figure 9.37 Semi truck fifth wheel suitable for biobased grease application

References

1. Pine Chemical Association. Retrieved 18 April 2010 from: http://www.pinechemicals.org/index.php?src=directory&view=Document_Repository_Public&submenu=Document_Repository.
2. Fatty Acid Profiles of Tallow, Fat, and Lard. Retrieved 18 April 2010 from: http://www.scientificpsychic.com/fitness/fattyacids1.html.
3. Hammond, E.G., and Lee, I. (1992) United States Patent #5089403 Process for enzymatic hydrolysis of fatty triglycerides with oat caryopses.
4. Zhang, R. (2009) Novel high performance calcium oleate complex grease. Presentation at the 21st ELGI Annual General Meeting 25–28 April 2009, Gothenburg, Sweden.
5. NLGI Annual Survey. NLGI Spokesman, National Lubricating Grease Institute 2009 Annual Survey Report, NLGI 2010.
6. Baart, P. (2009) Grease lubrication in radial lip seals. Luleå University of Technology. Department of Applied Physics and Mechanical Engineering Division of Machine Elements. Tryck: Universitetstryckeriet, Luleå, ISSN: 1402-1757 ISBN 978-91-86233-50-1 Luleå – www.ltu.se.
7. Zhu, P. (2009), "*Revolutionizing the Grease-Making Process*", Compoundings, pp.19–20, Vol. 59, No. 1.8. Madius, C. and Smet, W. (no date) *Grease Fundamentals: Covering the basic of lubricating grease*, Axel Christiersson, the Netherlands, www.AXELCH.com.
8. Madius, C. and Smet, W. (no date) *Grease Fundamentals: Covering the basic of lubricating grease*, Axel Christiersson, the Netherlands, www.AXELCH.com.
9. Wittse A.C. Carley, Method for continuous manufacture of high dropping point lithium grease. Patent no. 4444669, issued on 04/24/1984.
10. James Spagnoli *et al.* in a US patent no. 20050082014. OR You can sue the following: Patent #20050082014"; "Method and equipment for making a complex lithium grease"; Spagnoli, James E. (Mount Laurel, NJ, US), Nadasdi, Todd Timothy (Mount Laurel, NJ, US), Stober, Spencer Troy (Woodbury, NJ, US), Doner, John Phillips (Sewell, NJ, US), Graham, John Kenneth (Ardmore, PA, US), Sundstrom, Norman Charles (Yardley, PA, US), Grives, Paul Richard (Lumberton, TX, US), Carfolite, Barbara Anne (Wallingford, PA, US), Van Auken, James Fredrick (Beaumont, TX, US)", " April 2005", http://www.freepatentsonline.com/20050082014.html
11. Honary, L.A. (2009) New Developments in Biobased and Conventional Grease Manufacturing Processes, in *Eurogrease*, European Lubricating Grease Institute (ELGI), May 2009.
12. Metaxas, A.C. and Meredith, R.J. (reprint 2008) "Industrial Microwave Heating", Copyright: The Institution of Engineering and Technology, UK.
13. National Lubricating Grease Institute, Lubricating Grease Guide, Kansas City, MO, 1984.
14. Lou A.T. (2006) "Soybean Oil-Based Solid Stick Lubricant" in NLGI Spokesman, Volume 69, Number 12, pp 19–24.

10

Factors Affecting the Environment

10.1 Biodegradable and Biobased

Vegetable oils are in general considered biodegradable and meet the biodegradability standards of the American Standard for Testing and Materials (ASTM) and the Organization for Economic Cooperation and Development (OECD). Typical biodegradability results for base oils are shown in Table 10.1.

In the United States, vegetable oil-based lubricants fall under the definition of Biobased as defined by the United States Department of Agriculture (USDA). Discussions of biobased materials must first draw the distinction between "biodegradable" and "biobased". For biodegradable, two widely used designations are "readily" and "inherently" biodegradable.

- According to the OECD, the pass levels for a readily biodegradable product, there must be 70% removal of dissolved organic carbon (DOC) and 60% removal of theoretical oxygen demand and theoretical carbon dioxide within 28 days (OECD 301 Guideline for Testing of Chemicals).
- Inherently biodegradable defines no timing or degree of biodegradability (it will eventually biodegrade).

The biodegradability of mineral oil and mineral oil-derived base oils are reported in the literature [1].

To avoid controversies of the term biodegradable, the USDA introduced the term "biobased," initially intended as a label for products consisting of 51% or more "bio" materials. Later, USDA revised the approach to identify products based on their percentage of renewable materials. A broader definition of both "biobased" and "biodegradable" were provided by the author as follows.

- **Biobased:** Products containing natural renewable biological content like those made with agricultural materials (these products could be biobased and biodegradable).
- **Biodegradable:** Products that meet US (ASTM) or European (OECD) biodegradability requirements and could be made of biodegradable and or biobased materials.

Biobased Lubricants and Greases: Technology and Products, First Edition. Lou A.T. Honary and Erwin Richter

Table 10.1 Typical biodegradability results for base oils [1]

Mineral oil type	Amount biodegraded
White mineral based oil	25–45%
Natural and vegetable oil	70–100%
PAO	5–30%
Polyether	0–25%
Polyisobutylene (PIB)	0–25%
Phthalate and trimellitate esters	5–80%
Polyols and diesters	55–100%

These definitions make a more clear distinction between the terms biobased and biodegradable. It is essential to understand these terms, as they are not necessarily interchangeable. For example, there are some synthetic esters that are petroleum-derived and are biodegradable (meet the standards of biodegradability) but are not biobased. In the case of lubricants and greases, they are tested for their carbon 14 content to determine what percentage of their content comes from renewable carbon (bio) or what percentage of carbon comes from fossilized materials; this determines whether they are biobased. USDA now relies on ASTM D 6866-04 standards using radiocarbon and isotope ratio mass spectrometry analysis in order to determine the biobased content of these items. These test methods look at content alone and do not address environmental impact, functionality, or product performance.

Biodegradability of lubricant products refers to the bacterial breakdown of the oil and its additives over a period of time as measured in standard test methods. In the environment and exposed to the elements, all materials eventually break down and degrade into benign byproducts. Tests of biodegradability are used to determine the speed and degree of degradation under standard conditions.

The European countries established the Ecolabel through the European Commission in 2005 using a standard logo for identification (Figure 10.1).

Figure 10.1 European Ecolabel symbol [2]

Table 10.2 Requirements to use Ecolabel

	Greases current criteria	Proposed revision as of 2009
Renewability in % of weight	>=45	>=45
Biodegradation/accumulation in % wt.		
ultimately aerobic biodegradation (A)	>=75	>=75
inherently aerobic biodegradation (B)	<=20	}<=25
non-biodegradation. and non-bioaccumulation (C)	<=10	
non-biodegradation and bioaccumulation (X)	0	0
Aquatic toxicity in wt. %		
not toxic (D)	no limit	no limit
harmful (E)	<=25	<=25
toxic (F)	<=1	<=1
very toxic (G)	<=0.1	<=0.1
OSPAR listed components	not allowed	not allowed
Risk-phrase (ecotoxicity/human toxicity) for final product	not allowed	not allowed
Min. technical performance	"fit for purpose"	"fit for purpose"

There are certain requirements for the products to meet in order to qualify for using the Ecolabel. These, as explained by Nehls [3], are shown in Table 10.2.

The European norm EN 13 432 defines the characteristics of a material to be claimed as "compostable" and, therefore, recycled through composting of organic solid waste. According to EN 13 432, those characteristics are:

- For compostable materials biodegradability, namely the capability of the compostable material to be converted into CO_2 under the action of microorganisms. This property is measured with a laboratory standard test method: the EN 14 046 (also published as ISO 14 855: biodegradability under controlled composting conditions). In order to show complete biodegradability, a biodegradation level of at least 90% must be reached in less than 6 months. (Note: measurement errors and biomass production are experimental factors which can make it difficult to reach 100%: this is why threshold is set at 90% rather than at 100%.)
- Fragmentation and loss of visibility in the final compost (absence of visual pollution). Measured in a pilot scale composting test (EN 14 045). Specimens of the test material are composted with biowaste for 3 months. The final compost is then screened with a 2 mm sieve. The mass of test material residues with dimensions >2 mm shall be less than 10% of the original mass. (Note: also in this case a 10% tolerance is allowed, taking into account the typical error found in biological analysis.)
- Absence of negative effects on the composting process. Verified with the pilot scale composting test.
- Low levels of heavy metals (below given maximum values) and absence of negative effects on the final compost (i.e., reduction of the agronomic value and presence of ecotoxicological effects on the plant growth). A plant growth test (modified OECD 208) and other physico-chemical analyses are applied on compost where degradation of test material has happened.

Each of these points is needed for the definition of compostability but it is not sufficient alone. For example, a biodegradable material is not necessarily compostable, because it must also disintegrate during the composting cycle. On the other hand, a material that breaks during composting into microscopic pieces which are then not fully biodegradable is also not compostable.

The known and used standards of biodegradability for lubricants include ASTM D 6006 "Guide for Assessing Biodegradability of Hydraulic Fluids" and OECD 301 series tests. Both of these tests principally involve the same process with slight differences in reporting. The main concept is to monitor the conversion of the product to be tested for biodegradability to CO_2 during incubation in a controlled environment. The controlled environment could be aerobic or anaerobic and include presence or lack of light. Kopeliovich explains biodegradation as a process of chemical breakdown or transformation of a substance caused by micro organisms (bacteria, fungi) or their enzymes [4].

There are two types of biodegradation based on the extent of breakdown in a substance:

- **Primary biodegradation:** modification of some physical and chemical properties of the substance caused by activity of micro organisms.
- **Ultimate biodegradation:** total utilization of the substance resulted in its conversion into carbon dioxide (CO_2), methane (CH_4), water (H_2O), mineral salts, and microbial cellular constituents (biomass).

To test the biodegradability of materials, including lubricants, an electrolytic respirometer may be used. The test may be performed exposed to light as seen in Figure 10.2 (left), or covered to prevent exposure to light as seen on right.

10.2 REACH

Created by the European Parliament and Council in 2006, the new chemical regulation called REACH has created a new opportunity to reduce the damage chemicals may do to the environment. The new Registration, Evaluation and Authorization of Chemicals (REACH) provides a great opportunity for the promotion and use of biobased products and sets a new environment for business in Europe.

In Europe, the REACH work involves, among many others, the industry organizations such as the European Chemical Industry association for Health, Safety and Environment (CONCAWE) and the European Chemical Industry Federation (CEIF). In the United States, most organizations dealing with lubricants or greases have been dedicating time and resources to help their members who do business with European countries manage the REACH requirements.

The REACH regulation became effective in June 2007 and is enforced in all member states of the European Union. It replaces over 40 different rules and directives regarding chemicals. Through REACH, the industry is give the opportunity, as experts in their field, to register all chemicals used in their products that do not affect human health and or the environment. Both the downstream users and the manufacturers of chemical products are made responsible for chemical products and the authorities focusing on substances with highest risk while monitoring and enforcing the rules.

Implementation of REACH has been in phases, requiring companies to submit a preregistration of all chemicals they manufacture or import into Europe and are more than one ton per year, by December 2008. The final registration is due to be completed by November 2018.

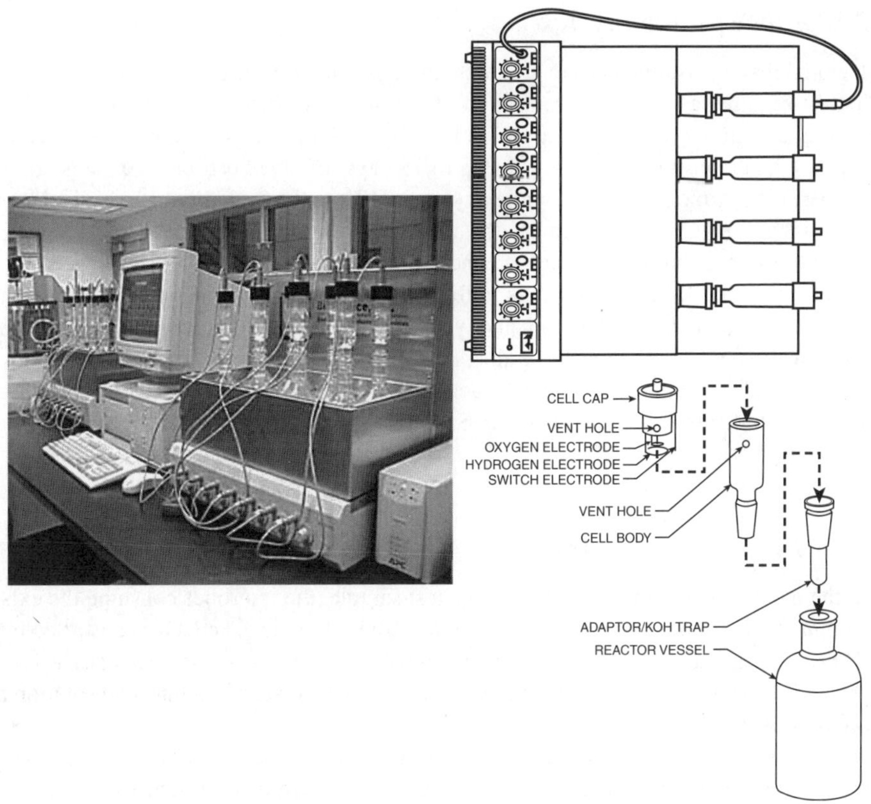

Figure 10.2 Electrolytic respirometer – and a view of a standards reactor assembly (right bottom). See Plate 32 for the color figure

However, chemicals whose volumes used exceed 1000 tonnes per year, require registration by the end of 2010.

The European Chemicals Agency (ECHA) is responsible for publishing the list of pre-registered substances. European companies have been able to act as agents or representatives for companies outside Europe to preregister substances.

The European Inventory of Existing Chemical Substances (EINECS) is a source of registered chemical substances in Europe similar to the American descriptive reference source, the Chemical Abstract Service (CAS). All chemicals including vegetable oils are registered by their EINECS and CAS that are helpful in developing safety data sheets for materials.

10.3 Biodegradation of Oils

Sharabi and Bartha [5] explain the process of measuring biodegradation as

> *. . . conversion to CO_2 upon incubation in aerobic soil is one of the standard test procedures to assess biodegradability. It may be measured with unlabeled test compounds in biometer flasks. In this case, the background CO_2 evolution by unamended soil is subtracted from the CO_2 evolution by the amended soil and the resulting net CO2 evolution becomes the measure of biodegradation.*

10.3.1 *Biodegradability Test*

Biodegradability measures the rate at which an organism consumes a test material.

The test requires a known concentration of test material to be placed into a closed flask containing a mineral medium and microbes; the solution is maintained at a constant temperature and continuously stirred for up to 28 days. Evolved carbon dioxide is absorbed by a potassium hydroxide solution, while oxygen is fed into the flasks at a at an equivalent rate. The amount of oxygen consumed by the inoculums during biodegradation of the test is measured through a computer, so at the conclusion of the test the biological oxygen demand (BOD) can be calculated, expressed as milligrams oxygen per milligram of test compound. Additionally calculated is the theoretical oxygen demand (ThOD), expressed in milligrams as the amount of oxygen required to oxidize a substance completely. Using these values the percentage Biodegradation of a sample is achieved. As an inoculum the test may use bacteria from sewage sludge from a domestic sewage-treatment plant, from natural water, soil bacteria, or their combination.

10.3.2 *Electrolytic Respirometer*

When the inoculums begin to metabolize the test sample, the microbes consume the existing oxygen inside the sample container and release carbon dioxide. The CO_2 is absorbed by the KOH trap, which causes a slight negative pressure in the sealed vessel. This slight vacuum triggers the electrolytic cell to supply oxygen via the electrolysis of a dilute acid solution until the pressure is equilibrated.

Primary biodegradation is measured by using an infrared spectrometer. Ultimate biodegradation is determined according to the evolution of CO_2 from the tested sample over that produced in a blank sample, which contains inoculum only.

Standard biodegradability tests:

- **ASTM D-5864 Standard Test Method for Determining Aerobic Aquatic Biodegradation of Lubricants.** The method is used for testing non-volatile oils, which are not inhibitory to the inoculum micro organisms.
- **CEC-L-33-A-94 of the Coordinating European Council (CEC).** The method is applicable for determination of primary biodegradability. It is widely used for testing engine *oils*.
- **OECD 301B, or Modified Sturm Test of the Organization for Economic Cooperation and Development (OECD).** The method determines only ultimate biodegradability by measuring evolving carbon dioxide.
- **OECD 301D or Closed Bottle Test of the Organization for Economic Cooperation and Development (OECD).** The method is used when the oxygen concentration in the test oil is not the limiting factor for degradation.
- **EPA 560/6-82-003 or Shake Flask Test of the US Environmental Protection Agency (EPA).**

The tests determine the rate of the biodegradation.

- **Readily biodegradable** – at least 60–70% (depending on the test type) of the sample oil is degraded.

- **Inherently biodegradable** – 20–60% of the sample oil is degraded.
- **Persistent** – less than 20% of the sample oil is degraded.

Table 10.3 shows the biodegradability requirements for ready and inherent biodegradability [1].

Table 10.3 Requirements for ready and inherent biodegradability

Title	Test parameter	Definition or pass criteria
Ready biodegradability		
301A: DOC die-away	% DOC removal	≥ 70% DOC removal within 28 a and within 10-d window after
301B: CO_2 evolution (modified Sturm test)	% CO_2 production	≥ 60% of theoretical CO_2 production within 28 d and within 10-d window after 10% CO2 production is reached
301C: MITI (I)	% BOD removal	≥ 60% theoretical BOD removal within 28 d
301D: Closed bottle	% BOD removal	≥ 60% theoretical BOD removal within 28 d and within 10- or 14-d window after 10% BOD removal is reached
301E: Modified OECD screening	% DOC removal	≥70% DOC removal within 10% 10-d window after DOC removal is reached
301F: Manometric respirometry	% BOD removal	≥ 60% theoretical BOD removal within 28 d and within 10-d window after 10% BOD removal is reached
Inherent biodegradability		
302A: Modified SCAS Test	% DOC removal in daily cycles	20–70% daily DOC removal during 12-wk testing
302B: Zahn–Wellens/EMPA	% DOC removal	20–70% DOC removal within 28 d
302C: Modified MITI Test (I)	% BOD removal and/or % loss of parent compound	20–70% BOD or parent compound removal within 28 d
304A: Inherent biodegradability in soil	Production of 14CO2 from radiolabeled substrate	Not specified
Simulation (confirmation)		
303A: Aerobic sewage treatment: coupled units test	% DOC removal	Degradation rate is calculated

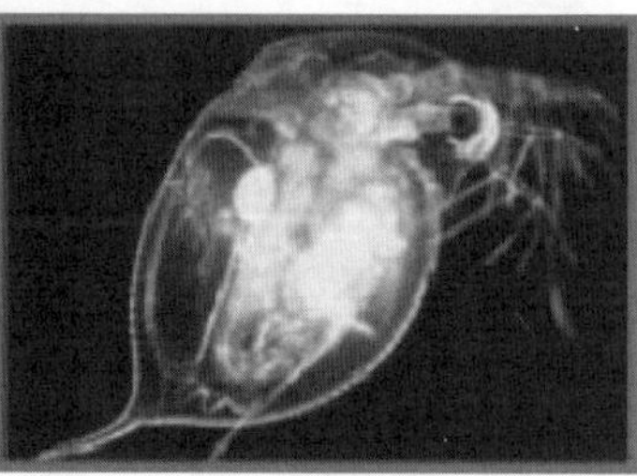

Figure 10.3 *Daphnia magna* can be grown and fed in conventional fish tanks (left) and actual photo of daphnia (right). See Plate 33 for the color figure

10.4 Toxicity Types and Testing Methods

Biobased fluids can be tested to determine if they present a health hazard when used as intended. Toxicity is measured according to the following:

a. **Aquatic toxicity:** The product's acute toxicity level of LC-50 is tested using ASTM D 6081 "Toxicity Studies – *Daphnia magna* for aquatic toxicity". Figure 10.3 shows *Daphnia magna*, which is a sensitive invertebrate used in one test for aquatic toxicity.

   Aquatic toxicity measures the effect a test material has on a population of aquatic organisms. The mortality of the populations is documented, and a lethal load percentage is calculated. To adapt this test to poorly water-soluble lubricants, the solutions are mechanically dispersed for 24 hours, and then the water accommodated fraction (WAF – for the portion of sample that is water soluble) is decanted, and this aliquot is tested for toxicity. Specific organisms (e.g. neonatal *Daphnia magna*; Figure 10.3) are then exposed to the various concentrations of sample WAFs, and the mortality rate of organisms is documented after 24 and 48 hours. A lethal load 50 is calculated, which is a statistically estimated loading rate of test material that is expected to be lethal to 50% of organisms.
b. **Toxicity:** water hazard. For a product to be considered non-hazardous to water, it should have a maximum rating of WGK 0, explained later.
c. **Toxicity, fish:** For the biobased product to be considered non-toxic it should have acute fish toxicity levels of LC-50 or greater than 1000 ppm for amounts measured in 48 hours.

Created in Germany, the ecotoxicity of materials is classified based on the degree of their impact on enlargement of water or *Wassergahrdungklasses* (WKG). Ecotoxicity is then classified based on the degree of water endangerment as WGK 0 to WGK 3, with 0 being non-endangering and 3 being more endangering.

10.5 Chronic Toxicity

Chronic toxicity is the ability of a substance to cause *long-term* effect on the impacted species. The ability of a chemical to cause harmful effects through *a single or short-term exposure* is considered acute toxicity. To measure acute toxicity a 50% lethal dose (LD50) which indicates that the product is lethal to 50% of the experimental living test specimens. This lethal dose rating is often followed by the method of administration and the affected species used for testing. For example, purified naphthenic oils may be rated as LD50/oral/rat <5 g/kg. This could be interpreted that based on the weight difference between laboratory rat and a human, a volume of $40\,\mathrm{l}^3$ of the oil would be needed to be consumed to reach an LD50.

Mutagenicity refers to changes caused in the DNA of the living organism. Typically, tests for mutagenicity (like the Ames test) are done on standardized micro organisms.

Other tests like carcinogenicity and terrestrial plant toxicity are used to determine the effect chemicals may have on the environment. Once the performance issues of biobased products are established, in general, these products can meet many of the minimum chemical safety standards with more ease than their conventional counterparts.

10.6 Terrestrial Plant Toxicity

Terrestrial toxicity measures the effect of sample substance on test organisms in soil. Seeds are planted in various concentrations of sample material blended in soil, and grown for twice the test organism's normal germination period. Seedling emergence, death rate, as well as shoot and root growth are documented during the test period. A plant growth chamber capable are providing controlled light, moisture and temperatures is used as per standards like OECD208 (Figure 10.4).

Figure 10.4 Plant growth chamber. See Plate 34 for the color figure

References

1. Harold, S. (1993) Biodegradability: Review of the Current Situation. The Lubrizol Corporation.
2. European Ecolabel symbol. http://ec.europa.eu/environment/ecolabel/ (Retrieved September 12, 2010).
3. Nehls (2009) Eurogrease. http://ec.europa.eu/environment/ecolabel/ecolabelled.../summary_modifs.pdf (Retrieved April 4, 2010).
4. Kopeliovich, D. http://www.substech.com/dokuwiki/doku.php?id=biodegradation_of_oils.
5. el-Din Sharabi, N. and Bartha, R. Department of Biochemistry and Microbiology, Cook College, Rutgers University, New Brunswick, New Jersey 08903-0231.

List of Useful Organizations

Name of Organization	Abbreviation	Location of Headquarters
American Chemical Society	ACS	Washington, D.C.
American National Standards Institute	ANSI	Washington, D.C.
American Petroleum Institute	API	Washington, D.C.
American Society for Testing and Materials	ASTM	West Conshohocken, PA
British Standards Institution	BSI	London, England
Bureau of Indian Standards	BIS	New Delhi, India
Canadian Fluid Power Association	CFPA	Toronto, ON
Canadian Standards Association	CSA	Mississauga, Ontario
Centers for Disease Control and Prevention	CDC	Atlanta, GA
Consumers International	CI	France
Defense Supply Center Columbus	DSCC	Columbus, OH
Environmental Protection Agency	EPA	Washington, D.C.
European Agency for Safety and Health at Work	EU-OSHA	Bilbao, Spain
European Association for the Coordination of Consumer Representation in Standardization	ANEC	Brussels, Belgium
European Committee for Standardization	CEN	Brussels, Belgium
European Telecommunications Standards Institute	ETSI	Sophia Antipolis,
Factory Mutual	FM	Johnston, RI
German Institute for Standardization	DIN	Berlin, Germany
International Energy Agency	IEA	Vienna, Austria
International Organization for Standards	ISO	Geneva, Switzerland

Biobased Lubricants and Greases: Technology and Products, First Edition. Lou A.T. Honary and Erwin Richter

Japanese Industrial Standards Committee	JISC	Tokyo, Japan
Japanese Standards Association	JSA	Tokyo, Japan
Mine Safety and Health Administration	MSHA	Arlington, VA
National Biodiesel Board	NBB	Jefferson City, MO
National Fire Protection Association	NFPA	Quincy, MA
National Fluid Power Association	NFPA	Milwaukee, WI
National Institute of Occupational and Safety Health	NIOSH	Washington, D.C.
National Institute of Standards and Technology	NIST	Gaithersburg, MD
National Lubricating Grease Institute	NLGI	Kansas City, MO
National Marine Manufacturers Association	NMMA	Chicago, IL
National Sanitation Foundation	NSF	Ann Arbor, MI
Nuclear Energy Agency	NEA	Paris, France
Occupationnal Safety and Health Administration	OSHA	Washington, D.C.
Organization for Economic and Co-operation and Development	OECD	Paris, France
Society of Automotive Engineers	SAE	Washington, D.C.
Standardization Administration of China	SAC	Beijing, China
Underwriters Labaratories	UL	Chicago, IL
U.S. Consumer Product Safety Commission	CPSC	Bethesda, MD
U.S. Food and Drug Administration	FDA	Silver Spring, MD
World Health Organization	WHO	Geneva, Switzerland

Useful Test Methods

American Oil Chemist's Society (AOCS)

- AOCS Da 15-48 (Wijs Method) – Standard Test Method for Determining Iodine Value.
- AOCS Cd 12b-92 – Oxidative Stability Index Note: Rancimat or Active Oxygen Method (AOCS Cd 12-57) could be used and converted to OSI values using AOCS approved conversion factors.
- **Polymers** (AOCS Cd 22-91). Usually degradation products in frying oil, polymers include dimers, trimers, tetramers, etc., and can be formed through oxidative and thermal reactions. The thin plastic like layer that forms in the frying pans is made up of polymeric materials. To test polymer levels high-pressure liquid chromatography is used.
- **Active Oxygen Method** (AOCS Cd 12-57) – Oxygen is bubbled through an oil or fat which is held at 36.5 °C (97.8 °F). Oil samples are withdrawn at regular intervals and the peroxide value (PV) is determined. The AOM is expressed in hours and is the length of time needed for the PV to reach a certain level. AOM is used as a specification for fats and oils. AOM hours tend to increase with the degree of saturation or hardness.
- **Peroxide Value (PV)** (AOCS 8b-90) – This is a test for measuring oxidation in fresh oils, and is highly sensitive to temperature. Peroxides are unstable radicals formed from triglycerides. A PV over 2 is an indicator that the product has a high rancidity potential and could fail on the shelf.
- **Rancimat** – Oxidative Stability by Rancimat – EN 14112 – Same principle of operation as Oxidative Stability Instrument; recently was approved as a part of Biodiesel standard test by ASTM (Method ASTM D 6751).

American Society For Testing And Materials (ASTM)

- ASTM D 5854 Practice for Mixing and Handling of Liquid Samples of Petroleum and Petroleum Products
- ASTM D 5864 Test Method for Determining Aerobic Aquatic Biodegradation of Lubricants or Their Components
- ASTM D 6006 Guide for Assessing Biodegradability of Hydraulic Fluids

Biobased Lubricants and Greases: Technology and Products, First Edition. Lou A.T. Honary and Erwin Richter

- ASTM D 6046 Classification of Hydraulic Fluids for Environmental Impact
- ASTM D 6081 Practice for Aquatic Toxicity Testing of Lubricants: Sample Preparation and Results Interpretation
- ASTM D 6139 Test Method for Determining the Aerobic Aquatic Biodegradation of Lubricants or Their Components Using the Gledhill Shake Flask
- ASTM D 6186 Test Method for Oxidation Induction Time of Lubricating Oils by Pressure Differential Scanning Calorimetry (PDSC)
- ASTM D 6278 Test Method for Shear Stability of Polymer Containing Fluids Using a European Diesel Injector Apparatus
- ASTM D 6304 Test Method for Determination of Water in Petroleum Products, Lubricating Oils, and Additives by Coulometric Karl Fischer Titration
- ASTM D 6351 Test Method for Determination of Low Temperature Fluidity and Appearance of Hydraulic Fluids
- ASTM D 6375 Test Method for Evaporation Loss of Lubricating Oils by Thermogravimetric Analyzer (TGA) Noack Method
- ASTM D 6546 Test Methods for and Suggested Limits for Determining Compatibility of Elastomer Seals for Industrial Hydraulic Fluid Applications
- ASTM D 6668 Test Method for Discrimination Between Flammability Ratings of $F = 0$ and $F = 1$
- ASTM D 93 Test Method for Flash Point by Penskey–Martens Closed Cup Tester
- ASTM D 92 Test Method for Flash and Fire Points by Cleveland Open Cup Tester (DoD adopted)
- ASTM D 5558 Test Methods for Determination of the Saponification Number of Fats and Oils
- ASTM D 97 Test Method for Determination of Pour Point of Petroleum Products (DoD adopted)
- ASTM D 130 Test Method for Corrosoveness to Copper from Petroleum Products by the Copper Strip Tarnish Test (DoD adopted)
- ASTM D 445 Kinematic Viscosity of Transparent and Opaque Liquids (the Calculation of Dynamic Viscosity), (DoD adopted)
- ASTM D 1500 Test Method for ASTM Color of Petroleum Products (ASTM Color Scale) (DoD adopted). Also, for biobased oils AOCS cc136-45 for Determining Color
- ASTM D 4057 Standard Practice for Manual Sampling of Petroleum and Petroleum Products (DoD adopted)
- ASTM D 4172 Test Method for Wear Preventive Characteristics of Lubricating Fluid (Four-Ball Method) (DoD adopted)
- ASTM D 4177 Standard Practice for Automatic Sampling of Petroleum and Petroleum Products (DoD adopted)
- ASTM D 4898 Test Method for Insoluble Contamination of Hydraulic Fluids by Gravimetric Analysis (DoD adopted)
- ASTM D 5185 Test Method for Determination of Additive Elements, Wear Metals, and Contaminants in Used Lubrication Oils and Determination of Selected Elements in Base Oils by Inductively Coupled Plasma Atomic Spectroscopy (ICP-AES) (DoD adopted)
- ASTM D 5306 Test Method for Linear Flame Propagation Rate of Lubricating Oils and Hydraulic Fluids
- ASTM D 5864 Test Method for Determination of Aerobic Aquatic Biodegradation of Lubricants or Their Components

- ASTM D 2271 Change in Viscosity in Pump Test – Based on the modified ASTM D 7043 now (formerly ASTM D 2271) revised for this specification using Eaton 20V Pump Cartridges and 10 gallons of sample oil instead of 5 gallons.
- ASTM D 2882 or Change in Viscosity in Pump Test – Based on the modified ASTM D 7043 now (formerly ASTM D 2271) revised for this specification using Eaton 20V Pump Cartridges and 10 gallons of sample oil instead of 5 gallons.
- ASTM D 2983 Test Method for Low-Temperature Viscosity of Lubricants Measured by Brookfield Viscometer.
- ASTM D 471 Test Method for Elastomer Compatibility of Lubricating Greases and Fluids. The tests involve immersing the sample material in the test fluid and specific temperatures for given periods of time and then checking the test specimen for:
  Change in Mass
  Change in Volume
  Change in Tensile Strength
  Change in Hardness
- ASTM D 1784 Test Method for Rust Protection by Metal Preservatives in the Humidity Cabinet. Used to measure rust preventative properties under high humidity conditions.
- ASTM D 4048 Test Method for Detection of Copper Corrosion from Lubricating Grease. A polished copper strip is immersed in a quantity of test product and heated to 100 °C for 24 hours. Strips are rated based on change in color against standards.
- ASTM D130 Test Method for Corrosiveness to Copper from Petroleum Products by Copper Strip Test.
- ASTM D 6973 Test Method for Indicating Wear Characteristics of Petroleum Hydraulic Fluids in a High Pressure Constant Volume Vane Pump. Dennison HF Series Tests – axial piston pump and vane pump
- ASTM D 5306 Standard Test Method for Linear Flame Propagation Rate of Lubricating Oils and Hydraulic Fluids

AFNOR

- AFNOR NF E48-692 Hydraulic Fluid Power. Fluids. Measurement of Filterability of HFC and HFD Class Fire Resistant Fluids

National Fluid Power Association (NFPA) Standards

Pump Tests

- VICKERS 35VQ-25 or latest replacement most notably ASTM D 6973
- DENISON High Pressure Pump Wear (p20, T6M) 973 – ASTM D 6973

MIL-PRF-Biobased-87257B

- ASTM D 5949 – Pour Point of Petroleum Products (Automatic Pressure Pulsing Method)
- ASTM D 2983 – Brookfield Viscosity
- NEW STANDARD – STORAGE STABILITY WITH OXYGEN BARRIER as described

- ASTM D 6304 – Determination of Water in Petroleum Products, Lubricating Oils, and Additives by Coulometric Karl Fisher Titration
- ASTM D 6793 – Determination of Isothermal Secant and Tangent Bulk Modulus

Society Of Automotive Engineers (SAE)

- SAE-AMS 3217/2B – Test Slabs, Acrylonitride, Butydiene (NBR-L), Low Acrylonitride, 65-75
- Vickers' 35-VQ-25 Test – Pump Performance Tests – One 50-hour tests, Pressure: 20.7 MPa (3000 psi), Temp. 65 °C for biobased oils, with pump speed of 2400 rpm, weight loss after 50 hours, Ring: 75 mg max, Vane: 15 mg max, Total: 90 mg max, along with visual inspection!

Federal Test Method

- FTM 791-3603 Seal Swell Test
- Flammability Test

Glossary

Additive: Any material added to base stock to change its properties, characteristics, or performance.

Anhydrous – as relating greases: A lubricating grease without water (as determined by ASTM D 128).

Aniline point: The lowest temperature at which equal volumes of aniline and hydrocarbon fuel or lubricant base stock are completely miscible. A measure of the aromatic content of a hydrocarbon blend, used to predict the solvency of a base stock or the cetane number of a distillate fuel.

Apparent viscosity: A measure of the viscosity of a non-Newtonian fluid under specified temperature and shear rate conditions.

Bactericide: Additive to inhibit bacterial growth in the aqueous component of fluids, preventing foul odors.

Bases: Compound that react with acids to form salts plus water. Alkalis are water-soluble bases, used in petroleum refining to remove acidic impurities. Oil-soluble bases are included in lubricating oil additives to neutralize acids formed during the combustion of fuel or oxidation of the lubricant.

Base number: The amount of acid (perchloric or hydrochloric) needed to neutralize all or part of a lubricant's basicity, expressed as KOH equivalents.

Base stock: The base fluid, usually a refined petroleum fraction or a selected synthetic material, into which additives are blended to produce finished lubricants.

Bleeding: Separation of liquid lubricant from a grease.

Blending: Blending is the process of mixing fluid lubricant components for the purpose of obtaining desired physical properties.

Boundary lubrication: Lubrication between two rubbing surfaces without the development of a full fluid lubricating film. It occurs under high loads and requires the use of antiwear or extreme-pressure (EP) additives to prevent metal-to-metal contact.

Bright stock: A heavy residual lubricant stock with low pour point, used in finished blends to provide good bearing film strength, prevent scuffing, and reduce oil consumption. Usually identified by its viscosity, SUS at 210 °F or cSt at 100 °C.

Biobased Lubricants and Greases: Technology and Products, First Edition. Lou A.T. Honary and Erwin Richter

Brookfield viscosity: Measure of apparent viscosity of a non-Newtonian fluid as determined by the Brookfield viscometer at a controlled temperature and shear rate.

Bulk appearance: Appearance of an undisturbed grease surface. Bulk appearance is described by:

- **Bleeding:** Free oil on the surface (or in the cracks of a cracked grease.)
- **Cracked:** Surface cracks.
- **Grainy:** Composed of small granules or lumps of constituent thickener.
- **Rough:** Composed of small irregularities.
- **Smooth:** Relatively free of irregularities.

Cetane number: A measure of the ignition quality of a diesel fuel, as determined in a standard single cylinder test engine, which measures ignition delay compared to primary reference fuels. The higher the cetane number, the easier a high-speed, direct-injection engine will start, and the less 'white smoking' and 'diesel knock' after start-up.

Cloud point: The temperature at which a cloud of wax crystals appears when a lubricant or distillate fuel is cooled under standard conditions. Indicates the tendency of the material to plug filters or small orifices under cold weather conditions.

Coefficient of friction: Coefficient of static friction is the ratio of the tangential force initiating sliding motion to the load perpendicular to that motion. Coefficient of kinetic friction (usually called coefficient of friction) is the ratio of the tangential force sustaining sliding motion at constant velocity to the load perpendicular to that motion.

Cohesion: Molecular attraction between grease particles contributing to its resistance to flow.

Complex soap: A soap crystal or fiber formed usually by cocrystallization of two or more compounds. Complex soaps can be a normal soap (such as metallic stearate or oleate), or incorporate a complexing agent which causes a change in grease characteristics – usually recognized by an increase in dropping point.

Consistency: The resistance of a lubricating grease to deformation under load. Usually indicated by ASTM Cone Penetration, ASTM D 217 (IP 50) or ASTM D 1403.

Copper strip corrosion: A qualitative measure of the tendency of a petroleum product to corrode pure copper.

Corrosion: The wearing away and/or pitting of a metal surface due to chemical attack.

Corrosion inhibitor: An additive that protects lubricated metal surfaces from chemical attack by water or other contaminants.

Demulsibility: A measure of the fluid's ability to separate from water.

Density: Mass per unit volume.

Dispersant: An additive that helps keep solid contaminants in a crankcase oil in colloidal suspension, preventing sludge and varnish deposits on engine parts. Usually nonmetallic ('ashless'), and used in combination with detergents.

Dropping point: The temperature at which grease becomes soft enough to form a drop and fall from the orifice of the test apparatus of ASTM D 566 (IP 132) and ASTM D 2265.

Dry film lubricant: A low shear- strength lubricant that shears in one particular plane within its crystal structure (such as graphite, molybdenum disulfide, and certain soaps).

Elastohydrodynamic lubrication (EHD): A lubricant regime characterized by high unit loads and high speeds in rolling elements where the mating parts deform elastically due to the incompressibility of the lubricant film under very high pressure.

Emulsifier: Additive that promotes the formation of a stable mixture, or emulsion, of oil and water.

Evaporation loss: The loss of a portion of a lubricant due to volatization (evaporation). Test methods include ASTM D 972 and ASTM D 2595.

Extreme pressure property: That property of a grease that, under high applied loads, reduces scuffing, scoring, and seizure of contacting surfaces. Common laboratory tests are Timken OK Load (ASTM D 2509 and ASTM D 2782) and Four-ball Load Wear Index (ASTM D 2596 and ASTM D 2783).

Flash point: Minimum temperature at which a fluid will support instantaneous combustion (a flash) but before it will burn continuously (fire point). Flash point is an important indicator of the fire and explosion hazards associated with a petroleum product.

Friction: Resistance to motion of one object over another. Friction depends on the smoothness of the contacting surfaces, as well as the force with which they are pressed together.

Fretting: Wear characterized by the removal of fine particles from mating surfaces. Fretting is caused by vibratory or oscillatory motion of limited amplitude between contacting surfaces.

Fuel ethanol: Ethanol (ethyl alcohol, C_2H_5OH) with impurities, including water but excluding denaturants.

Homogenization: The intimate mixing of grease to produce a uniform dispersion of components.

Hydrolytic stability: Ability of additives and certain synthetic lubricants to resist chemical decomposition (hydrolysis) in the presence of water.

Kinematic viscosity: Measure of a fluid's resistance to flow under gravity at a specific temperature (usually 40 °C or 100 °C).

Lubricating grease: A solid to semifluid dispersion of a thickening agent in liquid lubricant containing additives (if used) to impart special properties.

Naphthenic: A type of petroleum fluid derived from naphthenic crude oil, containing a high proportion of closed-ring methylene groups.

Neutralization number: A measure of the acidity or alkalinity of an oil. The number is the mass in milligrams of the amount of acid (HC1) or base (KOH) required to neutralize 1 gram of oil.

Neutral oil: The basis of most commonly used automotive and diesel lubricants, they are light overhead cuts from vacuum distillation.

Newtonian behavior: A lubricant exhibits Newtonian behavior if its shear rate is directly proportional to the shear stress. This constant proportion is the viscosity of the liquid.

Newtonian flow: Occurs in a liquid system where the rate of shear is directly proportional to the shearing force. When shear rate is not directly proportional to the shearing force, flow is non-Newtonian.

NLGI number: A scale for comparing the consistency (hardness) range of greases (numbers are in order of increasing consistency). Based on the ASTM D 217 worked penetration at 25 °C (77 °F).

Non-Newtonian behavior: The property of some fluids and many plastic solids (including grease), of exhibiting a variable relationship between shear stress and shear rate.

Non-soap thickener: Specially treated or synthetic materials (not including metallic soaps) dispersed in liquid lubricants to form greases. Sometimes called synthetic thickener, inorganic thickener, or organic thickener.

Oxidation: Occurs when oxygen attacks petroleum fluids. The process is accelerated by heat, light, metal catalysts, and the presence of water, acids, or solid contaminants. It leads to increased viscosity and deposit formation.

Oxidation inhibitor: Substance added in small quantities to a petroleum product to increase its oxidation resistance, thereby lengthening its service or storage life; also called antioxidant.

Oxidation stability: Resistance of a petroleum product to oxidation and, therefore, a measure of its stability.

Paraffinic: A type of petroleum fluid derived from paraffinic crude oil and containing a high proportion of straight chain saturated hydrocarbons; often susceptible to cold-flow problems.

Poise: Measurement unit of a fluid's resistance to flow, that is, viscosity, defined by the shear stress (in dynes per square centimeter) required to move one layer of fluid along another over a total layer thickness of 1 cm at a velocity of 1 cm/s. This viscosity is independent of fluid density and directly related to flow resistance.

$$\begin{aligned}\text{Viscosity} &= \frac{\text{shear stress}}{\text{shear rate}} \\ &= \frac{\mu\text{N/cm}^2}{\text{cm/s/cm}} \\ &= \frac{\mu\text{N/cm}^2}{/\text{s}} = 0.1\ \text{Pa s}\end{aligned}$$

Pour point: An indicator of the ability of an oil or distillate fuel to flow at cold operating temperatures. It is the lowest temperature at which the fluid will flow when cooled under prescribed conditions.

Pour point depressant: Additive used to lower the pour point or low-temperature fluidity of a petroleum product.

Pumpability: The low temperature, low shear stress-shear rate viscosity characteristics of an oil that permit satisfactory flow to and from the engine oil pump and subsequent lubrication of moving components.

Rheology: The deformation and/or flow characteristics of grease in terms of stress, strain, temperature, and time (commonly measured by penetration and apparent viscosity).

Rust preventative: Compound for coating metal surfaces with a film that protects against rust. Commonly used to preserve equipment in storage.

Saponification: The formation of a metallic salt (soap) due to the interaction of fatty acids, fats, or esters generally with an alkali.

Sludge: A thick, dark residue, normally of mayonnaise consistency, that accumulates on non-moving engine interior surfaces. Generally removable by wiping unless baked to a carbonaceous consistency, its formation is associated with overloading of the lubricant with insoluble substances.

Stoke (St): Kinematic measurement of a fluid's resistance to flow defined by the ratio of the fluid's dynamic viscosity to its density.

Synthetic lubricant: Lubricating fluid made by chemically reacting materials of a specific chemical composition to produce a compound with planned and predictable properties.

Texture: The texture of a grease is observed when a small portion of it is pressed together and then slowly drawn apart. Texture can be described as:

- **Brittle:** ruptures or crumbles when compressed.
- **Buttery:** separates in short peaks with no visible fibers.
- **Long fibers:** stretches or strings out into a single bundle of fibers.
- **Resilient:** withstands a moderate compression without permanent deformation or rupture.
- **Short fiber:** short break-off with evidence of fibers.
- **Stringy:** stretches or strings out into long fine threads, but with no evidence of fiber structure.

Thickener: The structure within a grease of extremely small, uniformly dispersed particles in which the liquid is held by surface tension and/or other internal forces.

Tribology: Science of the interactions between surfaces moving relative to each other, including the study of lubrication, friction, and wear.

Viscosity: A measure of a fluid's resistance to flow.

Viscosity index: Relationship of viscosity to temperature of a fluid. High viscosity index fluids tend to display less change in viscosity with temperature than low viscosity index fluids.

Viscosity modifier: Lubricant additive, usually a high molecular weight polymer, that reduces the tendency of an oil's viscosity to change with temperature.

Water resistance: The resistance of a lubricating grease to adverse effects due to the addition of water to the lubricant system. Water resistance is described in terms of resistance to washout due to submersion (see ASTM D 1264) or spray (see ASTM D 4049), absorption characteristics, and corrosion resistance (see ASTM D 1743).

White oil: Highly refined lubricant stock used for specialty applications such as cosmetics and medicines.

Yield: The amount of grease (of a given consistency) that can be produced from a specific amount of thickening agent; as yield increases, percent thickener decreases.

Index

Biobased Lubricants and Greases: Technology and Products, First Edition. Lou A.T. Honary and Erwin Richter